The Human Pedigree

'*I can assure you, doctor, that there's nothing congenital
on my side of the family.*'

THE HUMAN PEDIGREE

INHERITANCE AND
THE GENETICS OF MANKIND

Anthony Smith

LONDON: GEORGE ALLEN & UNWIN LTD
RUSKIN HOUSE   MUSEUM STREET

First published in 1975

© George Allen & Unwin Ltd 1975

ISBN 0 04 575019 X

Printed in Great Britain
in 12 point Fournier type
by The Aldine Press, Letchworth, Herts

Dedicated to
Adam
Polly
Laura

Contents

CONTENTS

Free service – Governmental involvement – Vasectomy –
Future ideas – Sperm banks – Evolutionary significance –
Medical involvement – The right to prevent

CONTENTS

Introduction

An explanation – General genetics – Immediate prospects –
The murder of a mongol – Tay-Sachs and the Jews – Double
recessive – The genetical problem

'*We hold these truths to be self-evident, that all men are created
equal*'

THOMAS JEFFERSON AND THE AMERICAN
DECLARATION OF INDEPENDENCE

'*The progress of biology in the next century will lead to a
recognition of the innate inequality of man. This is today most
obviously visible in the United States*'

J. B. S. HALDANE

'*The real content of the proletarian demand for equality is
the demand for the abolition of classes. Any demand for
equality which goes beyond that, of necessity passes into
absurdity*'

FRIEDRICH ENGELS

'*Heredity is the last of the fates, and the most terrible*'

OSCAR WILDE

This book is about human genetics. The subject is both huge and very
small; huge because it embraces the physical creation of every one of us,
small because nothing is known about the great bulk of our inheritance.
As a science genetics has flourished only recently; as an *ad hoc* procedure
for breeding better plants and animals it has a history stretching back to
the first efforts in this crucial direction.

Human genetics is looking in the pram to detect likenesses. It is disease,
like haemophilia. It is marriage laws, and incest, and the assortative mating
that we employ. It is twinning, and blood groups, and disputed paternity.
It is the choosing of sperm and eggs for the artificial aids to pregnancy.
It is sickle-cell anaemia, the Hapsburg jaw, sexual selection, Thomas
Malthus, vasectomy, amniocentesis, Morgan and *Drosophila*, Mendel and
1:2:1. It is also current evolution. Who is creating the next generation,
and who will not or who can not, by infertility or death or accident, leave

their genetic mark upon posterity? Human genetics is us. More important still it is those who will come after us.

It is also abortion, because this is doubly critical, killing 30 million a year and undoubtedly affecting our moral code. Will we become yet more prone to interfere, with selective abortion culling either the diseased or even the unwanted sex? Will we leave the decision until birth, rejecting the malformed or even the unappealing? It may be callous to recommend death for a hopeless offspring and to suggest that parents should try again but they, their welfare, happiness and finances, could profit immeasurably. Legal abortion rendered a hidden wish to surface. Much of modern biology will permit other aspects of human choice to manifest themselves, for good or for ill.

The world is obsessed with the quantity of its people and somewhat less concerned with their quality. The science of eugenics became a false science between the World Wars, taking wrong turnings in Germany (with Lebensborn and the Ahnenpass), in America (with the racial immigration laws) and in the thinking of thousands. Eugenics only means a genetic betterment, and there are now better ways of implementing its aims. There is no point in the birth of infants devoid of brain; there is no value in any other equally miserable ill-fortune. If counselling can reduce their incidence so much the better for counselling.

Within this book the purpose is to present new facts and the new problems they bring with them. The simplicity of old standards has gone for ever. The medical profession is being forced into new decisions on our behalf but is also reluctant to yield up its role as moral arbiter. The legal profession is being cautious in altering old laws to cope with current change. How remarkable, for example, that an AID child is classed as illegitimate and has fewer rights than any adopted infant. The rest of us will have to make the pace.

The logic behind these future pages is to introduce a range of topics that impinge upon human genetics, thereby enabling possible conclusions to be founded upon a certain breadth of information. It *is* relevant how they bred dogs into today's variety, how Bakewell made his cattle, how the *Bounty* men fared on Pitcairn, and what Bartolommeo did to re-create another Leonardo. It is important to look at mankind's evolution, at race, at genetic truths and their converse. The current debate on ethnic intelligence is probably less revealing but an extraordinary story, all the same. So too porphyria in South Africa, Tay-Sachs in American Jewry, and the origin of smallpox not so very long ago.

The actual details of man's inheritance hopefully become more intriguing in this wide-ranging context. Six fingers, baldness, myopia, stature, idiocy, ear-wax, dwarfism, digital hair, naevi, deafness, albinism, diabetes, epilepsy – there seems no end to the examined oddities of heredity. But reading of fraught genetical problems, or so I believe, makes any decision easier about them. Should the deaf marry each other, as they tend to do, thereby compounding their defective genes? Is it worth the effort and money to find the phenylketonurics? Should we have a different and more Spartan attitude to spina bifida?

There are some personal idiosyncracies in this book needing explanation. Strict scientific writing is full of qualification and keeps halting in its flow; general scientific writing, or so I believe, should be smoother, more ready with a generalization, less willing to stop and define at every turn. To assist with this intent there are two glossaries, one about people and one about words. To have always to give a person's academic rank when he is quoted, and to have to outline a disease merely because it has been mentioned, can be extremely inhibiting to any text.

Another personal prejudice is the attempt to restrict my own opinion. Of course it is there, as all facts speak through an interpreter, but it is subdued; there is no enthusiasm for nailing colours to the mast at every turn, for referring to *vile* experiments, to *idiotic* notions, even to *horrific* atom bombs (as if there were any other kind). Besides, not everyone emerges with the same viewpoints after encountering the same facts, and this is a book of facts far more than views. Ethics, as Bertrand Russell said, should be founded on an understanding of the truth, not vice versa.

The business of breeding is of creating, yet again, a further array of biological differences. These, in themselves, need not create injustice. What constitutes an injustice is the emotion aroused by an inequality, not the inequality itself. If we could give up our idea of superiority and inferiority we might start to appreciate our differences for the colourful things they are. As Eliot Slater wrote: 'Biological differences are basic; what we need to do is to adjust our world to make the utmost of them, and to enjoy them.' This book, then, is an enjoyment of those differences.

From the outset there is dilemma. We strive for equality of opportunity, but know that we are born unequal. We struggle for freedom, for the individual to do as he pleases, but impose restrictions upon society for the greater good of all. We feel that each man and each woman should marry whom they wish, breed as they please, and create another generation by this form of wilful chance; but we also wish to restrict numbers, to curtail hopeless imbeciles, to hand down fewer problems to our descendants. If we had remained as wasteful predators, stampeding bison from cliff tops and assuaging short-term appetites, we would have remained as primitives numbering a few millions at most. Instead we chose to domesticate animals, to husband resources, to select and breed our requirements, to improve – from our point of view – upon the wild strains; but at the same time we have applied little or none of this to ourselves. We have bred randomly. We have not mated with forethought or with consideration for the products of our relationships.

The conflict of human genetics is now coming to the surface. The current enthusiasm for equality and for improving the environment of all our lives has led to considerable rejection of the idea that there is fundamental inequality. To suggest, as some have done, that intelligence is 80% inherited is to be attacked with violence. To imply that the environment is everything, and inheritance is trifling, is to be met today with the kind of applause normally reserved for political polemics. As a society we attempt to lavish all possible care upon the newborn child, to nourish its mind and body, to safeguard its rights and virtues; but we are loath to interfere with its conception. To speak of eugenics is instantly to be reminded of (and probably to be appalled by) much done in its name, of the assertion that the fecund poor were producing second-rate citizens soon to swamp the whole.

Genetics, when it arose as a science (the word being coined in the earliest years of this century) was hailed much as a new drug is hailed. It was a cure-all. Pragmatically it had achieved fantastic results in the breeding of crops and animals; scientifically it would solve mankind's major ills. It would rid the world of inherited deficiency, of drunkenness and lawlessness, idleness and stupidity. It would even banish poverty if only by preventing the poor, those people of such inferiority that they cannot be other than on the bottom rung of every ladder. All geneticists (more or less, but notably in the United States) became eugenicists, and many eugenicists moved over into politics, the better to advance their various enthusiasms. There was considerable talk of the Nordic race, the particular grouping of mankind that had done so much to subdue the planet. There was heady argument, notably from this same band of people, that other groups were not only less adequate but less estimable as migrants, less suitable as breeders and less worthy even to live.

As a result the science of genetics became tarred with some of this evil.

Its findings were thought suitable for animals but irrelevant or unwarranted for human beings. A bishop wrote to *The Times* (of London) shortly after World War II, when the possibilities of science had suddenly reached a new peak, that all nations should agree to destroy the formula of the atomic bomb. 'There is no simple remedy for ignorance so abysmal,' said Peter Medawar afterwards. Similarly there is no remedy for wishing to deny the findings of genetics. A lust for political equality is one wish, and need not have as its bedfellow a deprecation of all fact asserting physical or mental inequality.

Eugenics went wrong when its promoters put politics before science, and it may be necessary to feel more wary of politicians than scientists. Certainly Medawar thinks so: 'I have more faith in scientific than in political man if only because the solution of the scientific element of the problem is much easier, being so much less handicapped by the inertia of bigotry and self-interest . . . There must be very few wicked scientists. There are, however, plenty of wicked philosophers, wicked priests, and wicked politicians.' Human genetics cannot of itself be wicked; nor, as the rest of us hold the strings to legislation, can human geneticists.

The science has even been condemned for its potentialities. There is frequent mention of clones, of factory incubation, and of sterilization of all but a few selected studs. Such practices may have become feasible, or will shortly be so, but there is total doubt about their likelihood. What kind of society would permit cloning, the production of identical individuals? What male ego would happily welcome alien sperm? There have been tremendous changes in our manner of life, but there are still predilections for the family, for having one's own children, for the differences inherent in mankind.

Similarly, to talk of test-tube babies is to imply a simple alternative to the natural method of uterine incubation. Can we really imagine that women will wish to give up, or be forced to give up, their pregnancies? If so, what about the cost? To keep a patient in hospital, even without drugs or treatment, is expensive, say £60 a week. To keep a premature baby in an incubator, bathed in oxygen, cleaned with difficulty, and fed repeatedly, is yet more expensive, say £200 a week. To maintain a foetus throughout thirty-eight weeks, to supply it with nutriment, to preserve its temperature, to remove unwelcome metabolites from its amniotic fluid – all of this could average out at £200 a week, or £7,600 per incubation. The technique will surely become possible, but will it become acceptable either by women (or men) or the state? Finally, will normal couples readily give up their ability to become parents, merely because one authority has decided that only an élite should provide the germplasm for the next generation?

Some prospects of reproductive biology are less distant, such as the ability to choose the sex of future offspring. Or to check on general foetal soundness via the probing of amniocentesis, and then to abort if unsatisfactory. Or to take sperm or eggs from donors. Or to check on a possible mate for signs of genetic disorder that could prove unwelcome. Or to learn that a particular union between two people would be disastrous for their progeny. The future is already here with some of these techniques. With the others, and with many more of similar advancement, their future is not embroiled in some distant culture quite foreign to our own but will be with us either tomorrow or very soon.

Only current prospects and current realities form the material for this book. The fact that, by and large, we are permitting abortion is not irrelevant to further forms of interference. We are taking contraception out of its medievalism of barriers and douches and bringing it into line with our scientific progress. We have genetic counsellors, and we will have more of them. We are manipulating our reproduction more than ever before, and will continue to do so, increasingly.

We are therefore confused. The suddenness of things has caught us, ethically, on the hop. We have no idea whether legalization of abortion enhances or lessens our respect for life. (Abolition of capital punishment occurred in many countries at precisely the same time as the acceptance of abortion.) We are even at a loss over words, resenting the fact that anyone should be called an idiot or an imbecile, and preferring labels such as backward, retarded or handicapped; but is an idiot merely afflicted with a handicap? We abort healthy foetuses by the million and then, so help us, keep even anencephalics alive for weeks on end despite their total lack of any brain. We like handing moral decisions to our doctors, and then dislike their unwillingness to do as we please. Above all, we have codes that we apply to others and a ready willingness to forget these strictures should we ourselves become enmeshed in them.

By way of further introduction, and of example, two recent cases help to demonstrate current uncertainties. Of course we should marry and mate whom we please, or so we say. Of course it is wrong to take the life of any individual once he or she has been born, or so we feel. Read these two stories, and then wonder if previous beliefs are quite so firmly rooted as before.

The first involves a murder. An American couple, the parents of two normal children, were in their mid-thirties when their third child was born. The new offspring was a mongol, being afflicted with that combination of malformations more correctly known as Down's syndrome. When the mother overheard this news about her child, she immediately expressed the opinion that she did not want the boy, but there was an additional factor. The child was suffering from a form of intestinal blockage known as duodenal atresia. This is a simple defect, and curable by surgery. The offending blockage has to be punctured, and then all is well.

As with all operations on children the parents were asked for their permission. The mother refused, and her husband supported her decision. It was explained to them that mongols were perennially happy, and frequently a great joy to all who knew them. Their average IQ was certainly low – between 50 and 80 – but this did not prevent them performing simple jobs. It was also explained that the blockage meant no food could leave the stomach. Therefore, until this barrier was removed, the child would starve. The parents remained adamant; they would not give their permission for the operation. It would cure the atresia, they argued, but not the mongolism. Faced with this refusal the hospital could have taken the case to court, and attempted to have the parental decision overruled. This procedure is adopted, for example, when parents refuse blood transfusions, as Jehovah's Witnesses are wont to do. However, in the case of the mongol neither did the doctors go to court nor did they operate. The defective infant was placed in a side room and died from starvation eleven days later.

It had apparently been felt by the physicians that a court would not have countermanded the parental wishes owing to the serious mental abnormality. If the court had done so it would have been imposing a financial and emotional burden upon the reluctant parents or upon an equally reluctant society. But one of the doctors said afterwards: 'Who has the right to decide for a child? The whole way we handle life and death is the reflection of the long-standing belief that children do not have any rights, that they are not citizens, that their parents can decide to kill them or let them live, as they choose.' Anyway the parents chose, the doctors did not intervene, and both they and the nursing staff felt it was 'clearly illegal' to hasten the baby's death by the use of drugs.

The mother gave her reason for refusal quite simply: 'It would have been unfair to the other children of the household to have raised them with a mongoloid.' The father, in giving his support to the refusal, said that his wife knew more about these things. The doctors said the very low IQ had been the crucial factor. Had the child been of normal intelligence they would have attempted to reverse the parental decision. The nursing staff were resentful over the way in which the child had been left, neither helped to life by medication nor helped to death. When the baby died, 'I was glad that it was over. It was an end for him,' said one of them.

Other countries or other hospitals might have reacted differently. In Britain a court order to go ahead with the operation might have been obtained, and the baby would then have been made a ward of court. Or the surgeons might have carried out the operation without waiting or even applying for such an order. The parents could then have sued, but it is expensive and difficult to sue the medical profession. In any case, had the baby been made to survive and had the parents won their subsequent legal action, it would still have been the task of the parents to look after their malformed and unwanted offspring. It needs a strong will either to refuse an operation or to sue when it has been performed. Most parents do nothing; they acquiesce in the situation that fate has presented to them.

So should doctors or parents decide? The admission of a mongol child into a family is an event of major consequence for the parents, for the siblings, for close friends and relatives, but not for the doctors. On what grounds do the courts decide? Is intelligence the key, with survival ordained for some limbless distortion of a child whose brain is well, but curtailed for some other infant, strong in body and weak in mind? The case can challenge rigid convictions. No longer are simplistic attitudes able to provide such a cosy shield with which to go through life. Thou shalt not kill is, as an injunction, about to be eroded. Thou shalt not kill except in special circumstances may well be taking its place. It would seem as if the current ethic is that thou shalt not kill any whose IQ is normal.

Now to a loathsome disease, and another dilemma. It questions our right to mate with the person of our choice and, by extrapolation, to mate and produce children without preliminary examination. Tay-Sachs disease is caused by a recessive gene. Any victim of it is unable to produce the enzyme hexosaminidase, a deficiency sufficient to cause death. Any affected baby appears normal at the time of birth but, after a few months, deteriorates progressively. Vision is affected, then the mind and then the body in general, and the child will die within two to three years. There is, as yet, no cure, nor the immediate likelihood of one.

We all possess harmful recessive genes, but they can only manifest their harmfulness when we possess two of a kind, one from each parent. To have a single gene is almost always unimportant, for its abnormality is masked by the normality of the gene that partners it. In the case of Tay-Sachs the normal gene will produce enough of the enzyme, and therefore all is well.

There are several distinguishing features about this disease. Firstly, it is possible to discover if you are a carrier and are possessing one abnormal gene. By the same token it is also possible for your potential mate to discover this. Secondly, the abnormal gene is extremely common among the Jewish population of America who are of eastern European origin. Why this should be so is unknown, but perhaps one affected individual or family produced an abnormally large number of descendants after arriving in the New World. Anyway the relative proportions both of the genes and of affected children among these Ashkenazi Jews make the issue a critical one for them.

In their community the ratio of abnormal to normal genes is 1 in 60. As everyone receives two of this kind of gene from their two parents, the ratio of carriers to normal people is therefore 1 in 30. This means, if one boy happens to be a carrier, that the chances of him wishing to marry a girl who is also a carrier are 29 to 1 against. In other words, the proportion of couples who are both carriers, who carry two abnormal genes between them, is 1 in 900. As only 1 in 4 of their children will receive the affected gene from each parent the proportion of affected children will be 1 in 3,600.

That ratio is much less than the incidence of mongolism in normal societies. It is far less than the incidence of severe congenital malformations of all kinds which, in many areas, run as high as 2, or 72% in every 3,600. Nevertheless the disease of Tay-Sachs, by singling out one particular group of people and by being identifiable in the carrier state, does pose new problems. There is nothing, as yet, that can be done about mongolism, save to breed earlier and reduce the number of older mothers, maternal age being closely correlated with its incidence. There is nothing, as yet, that can be done about the great mass of congenital malformations (16,000 a year in Britain alone; four times that number in the United States) but something can be done about Tay-Sachs. A man can check if he is a carrier and, if so, can then check his mate.

The girl may be everything the man could desire – before he discovers that she too is a carrier. To know in advance that the odds against each child receiving the double dose are only 3 to 1 is very alarming. To go ahead with the marriage, to admire the first-born child, to watch for signs of deterioration, and possibly to see it die, is to know that precisely the same hideous chain of events can happen the next time. And the next. To refuse the tests, which only involve the donation of a small amount of blood, and to be ignorant is to avoid a peace of mind that would otherwise be possible. To accept the tests, to discover the worst, and then to cancel that partnership for the sake of another may sound cold-blooded, but it must also be chilling to watch a child's death imposed by the narrow odds of 3:1.

Cost is relevant. The death of a Tay-Sachs victim is extremely expensive, adding up to about £10,000 at the end of a long stay in hospital. Whether this burden is paid for privately or publicly the argument for a screening programme remains the same. With an incidence of 1 in 3,600, and with that £10,000 cost per death, the screening would be economically sensible if the price per test were less than £2.75. This assumes, of course, that a positive result for a couple causes them to leave each other, or refrain from bearing children, or opt for artificial insemination from a non-carrier, or demand an abortion once, on average, in every four pregnancies. (The disease can be detected during the fourth month by the removal and examination of some amniotic fluid. There is still time for a simple therapeutic termination.) In any case, £2.75 is a small price to pay, either by the individual or the state, if it prevents the possibility of a £10,000 bill later on.

The argument will become increasingly important as more and more harmful genes become detectable. We all have them but are largely ignorant of our ownership. Should a couple produce, for example, an albino child they will know that both of them, husband and wife, possess this recessive and harmful gene. But they may also remain ignorant if the fates decide that their two recessives never happen to come together. Should conception be such a gamble? Should the tests, as they become available, be free for everyone? After all, in the lottery of finding a mate, a decision often has to be made between two or more contestants. To know that one of them shares, say, three harmful genes is to be forewarned that the chances of a malformed child are not 1 in 4 (or 25%), as with a single pair of recessives, but in the proportion of 37 in 64 (or 58%).

There were new issues in the mongol case. In effect the parents murdered the child and society then condoned their action by bringing no charge against either them or the doctors and nurses who had known perfectly well what was happening. The implication is clear that one law exists for the normally intelligent and another exists, or is wholly absent, for the mentally deficient. Is this what we want? With Tay-Sachs and other disorders of this nature (see the chapter on inheritance for some idea of their profusion) should there be a compulsory genetic test before marriage? If the results are unsatisfactory should marriage be forbidden or should counsel merely be proffered? Or should the marriage be permitted with only children being forbidden? The state is the current arbiter, almost everywhere, of who is permitted to marry whom, and will – in many cases – have to bear the burden of caring for any unfortunate products of conception. Therefore should the state become stricter, refusing cousin marriages, for example, and permitting only those that incur no degree of inbreeding?

The two cases, the one of selective rejection of a child at birth, the other of greater deliberation in the choosing of a mate, both indicate future trends – and the kind of material to be encountered in this book. The business of reproduction was once elementary, the rules clear cut, the medical ethic straight from Hippocrates. Children happened, then children died or children survived. Brand-new problems needing brand-new decisions are now being brought home to all of us and the doctors' dilemma is no longer only theirs. It is ours.

CHAPTER 1

Evolution of mankind

The haste of early man – *Homo sapiens* – Dark skins and
vitamins – Racial variation – Genetic diversity – Influence
and origin of disease – Migration and change in stature –
Hybrid vigour

'*Evolution is the most powerful and the most comprehensive
idea that has ever arisen on Earth*'

JULIAN HUXLEY

'*If one looks with a cold eye at the mess man has made of his
history, it is difficult to avoid the conclusion that he is afflicted
by some built-in mental disorder which drives him towards self-
destruction*'

ARTHUR KOESTLER

Genetics is the past, the present and each generation's permanent contribution to the future. The past reaches back to the very beginning, to the
first particle of organic matter capable of self-replication, and all biology
has an equally ancient ancestry. Mankind, of course, is no exception;
those several thousand million years are also our inheritance. It is as well
to contemplate that past before advancing to the present. Nevertheless
it would be a lengthy procedure to view more than a fraction of that
lineage, and a short glance merely at the recent human past will probably
provide sufficient provocation for thought about our genetic origins. Two
million years are, from our ephemeral mayfly viewpoint, sufficient to
ingest for the time being.

To visualize evolutionary change, where we can examine both the end
(ourselves) and the beginning (in the fossil record) is to experience a
sense of loss at the distance in years. The time scale is so hard to comprehend. Take *Homo habilis*, for example, the man whose immortalized frag-
ments were dug up with such devoted patience by Louis and Mary Leakey
in the gorge of Olduvai. The kind of people given this name, tool-makers
with a capable hand and a modern form of foot, lived two million years

22

ago. They walked upright. They had a jaw possessing sufficient space within it for a tongue to be used in speech. They had a good brain with a volume of 680 cc, or half the average brain size of modern man. The stature of these people was small – roughly 4 ft – and therefore smaller by a foot or so than the more primitive man-apes known as Australopithecines who lived both before them and, so some say, simultaneously.

The growth of the brain during its transition from *habilis* to modern man has been – to borrow a term from the palaeontologists – explosive. It was the biggest brain of its day and it has enlarged greatly since then. During the past two million years there have been only 80,000 generations of men, and during that time the brain has exploded by 800 cc, or one-hundredth of a cubic centimetre in every generation. As explosions go such increments are modest, but in evolutionary terms they are extraordinary.

Cell size and cell number in the human body are also difficult to appreciate, and perhaps it is better to think of the number of brain cells involved in that extraordinary growth. The modern brain contains fourteen thousand million cells within its 3 lb. of substance. Assuming a similar density for the brain of *habilis*, we today have 7,200 million more cells than he had, or a gain of 90,000 cells in each generation, or 10 cells a day. Suddenly the rate of growth does seem considerable.

The other easily measurable gain is that of stature. Size, one can safely assume, has not been so crucial in the evolution of man as the swelling of his cerebral hemispheres, but a growth from 4 ft to today's (European) average of 5 ft 7 in. is still impressive. It means a steady enlargement of 0.0002 in. in every generation. At this speed, and assuming constancy, the men at the time of Christ were, on average, 0.016 in. shorter than we are, or less than half a millimetre different. The change seems minute, but the palaeontologists, accustomed to many other pedigrees from the fossil history, are happy to call this growth extremely rapid. We can be surprised, when we peer into ancient burial grounds, whether Viking or early Egyptian, to find their remains so similar in stature to our own, but in our evolutionary history they are practically contemporaneous.

Habilis to modern man involved changes other than height and brain size. Teeth altered; the big canines were reduced so that tearing and swallowing became a matter more of cutting and chewing. The spine grew S-shaped and better for running. The heel increasingly took on its modern form. The jaw shrank in size. The *habilis* hand was fairly modern but its thumb had not rotated from the wrist and was therefore less effective

23

(try pinning the thumb at its base to prove the point). Therefore the *habilis*-made implements were little more than pebbles with some flakes struck off, to produce more angular objects. The precision grip – the delicate contact between thumb and finger, possible even when the strong grip with the ball of the hand is being maintained – had not yet acquired its fullest expression.

The brain reached its modern size roughly 80,000 years ago, but everything we think of as achievement has happened since it ceased its growth. Chipped-stone axes and flake tools did not reach the peak of their development until 40,000 years ago (and some say that true hunting did not begin until this time). The carvings and paintings of the Upper Palaeolithic only started some 30,000 years ago. The domestication of animals only began about 10,000 years ago, and was then followed by the amazing Neolithic revolution that set us most irrevocably upon our current path. The population boomed at that time, perhaps to 200 million, and such an increase was not to occur again until the start of industrial times.

This spate of recent advance has all been performed by the body and the brain of a savage. What else were we 80,000 years ago? In 1870 Alfred Russel Wallace was pointing out that the brain of a primitive man was very little inferior in biological terms to that of a philosopher, and quite disproportionate to his requirements. As Julian Huxley put it: 'Only negligible genetic changes in man's intelligence or other capacities have taken place since the Neolithic period, and quite possibly since Cromagnon times.' When the *Endeavour* sailed into an Australian inlet two centuries ago, busily applying science and technology from the western world to the unknown Pacific, it observed three Australian aborigines who neither stopped nor apparently glanced up from their fishing as the British bark sailed by. Both those on deck and those on shore were *Homo sapiens*, the same species with, more or less, the same physique and the same innate ability.

It must have been tempting for those on board to feel a cut above the black primitives. Someone handling a compass, a chart, a sextant, and dressed in a fine suit of cloth is only too ready to believe that he somehow is the creator of those things. The aborigines, naked and waiting only for the fish to bite, were certainly inferior in the range of their possessions, but all people of Europe, or of European descent, should remember that neither they nor their forefathers on that continent had a hand in inventing the clock or the calendar, weights or measures, or a means of recording results. Therefore, in this sense, the aborigines and the Europeans were equal, with only a different inheritance of culture between the two of them. It is easy to forget this point.

Genetic changes may have been negligible in recent millennia, as Huxley said, but he added that 'there have been almost incredible alterations in cultural apparatus and achievement, such as religion, art, science, law, technology, social organization, literature and education'. *Homo sapiens* took at least 50,000 years, after he had acquired his modern brain, before he managed to improve anything. During that period, much like those aborigines, he did well in the good times and fearfully in the bad. There was already the culture of some taught behaviour, such as fire, cooking and shelter, and suddenly there must have been much more when, with an apparent abruptness, the primitives took those strides that led to modern man. The hunter-gatherer became the domesticator, the agriculturalist, the villager, and the man we know today.

Of all the people who have ever lived since *Homo* first began two million years ago, of the estimated total of eighty thousand million humans who have lived out a life-span on this planet, at least 90% have been hunter-gatherers. Since then about 6% have been agriculturalists, and the final few per cent have lived with industry. In a mere second of cosmic time man has 'produced astonishing achievements' said Huxley, 'but he has also been guilty of unprecedented horrors and follies. And looked at in the long perspective of evolution he is singularly imperfect, still incapable of carrying out his planetary responsibilities in a satisfactory manner'. In short, and forgetting the culture change, he is still biologically a hunter-gatherer. That is how selection created him, and the few generations since then have not altered the basic form of his inheritance.

It is easy, with current conceit, to imagine that we today formed the objective of those two million years in the forest, on the plains, in the wilderness. That increase in brain size appears to trump the matter. We had a destiny, it would seem, to become the dominant type, to subdue – as Genesis instructs – the earth. In fact, we might have died out and perished, as more than 99% of all species have done, but we did not, and acting on our behalf must have been both our ability to adapt to a wide range of climates and our own considerable variety, our difference between groups and individuals.

Natural selection does not favour a systematic change in a fixed direction, as one might think from looking at *habilis*, then *erectus* who followed him, then *sapiens*, the last in the line. Mostly it is acting to remove the abnormal, the deviants from the mean, the very big, the very small, or even the unduly courageous, the fastest, the cleverest. Much selection is involved, to use a phrase of James F. Crow, in maintaining 'the genetic status quo'.

Nevertheless there would also have been considerable variability, among tribes, between tribes and within areas. Part of that variability is the visible difference between the major groupings of mankind, the races that are the cause of so much current friction. Instead of antagonism there should only be rejoicing at this considerable expression of divergence among our human kind.

It is as well to hesitate before embarking upon the subject of race. It may therefore be doubly valuable to discuss how at least one character came about that distinguishes tropical man from his temperate neighbour. Broadly speaking, as we know only too well, there are three groupings, the black, the white and the yellow (this colour code is hardly accurate but is in common use). The problem is to explain the darker tropical colour of black people when, as everyone appreciates who has worn dark or light clothes in strong sunshine, the whiter protection is cooler. It is of course important to maintain the correct and comfortable body temperature, but the prevailing theory on this subject implicates a far greater importance, that of vitamins. It would seem as if vitamin D, and not heat, is the crucial factor.

A strangeness of the body is that it is perfectly able to make hundreds and hundreds of different chemicals from the random assortment of foods given to it, but is quite incapable of manufacturing a dozen or so that are equally essential. There is no similarity between them, save that death in a variety of forms is inevitable if they are withheld. Vitamin D is necessary if calcium is to be processed to make bone. With absolutely no D, a difficult event if the person is exposed to any sunlight, the calcium process will break down. With insufficient D, a far too frequent event in impoverished and sunless communities such as the industrial slums of the last century, growing bones suffer and rickets is the result.

However, there are crucial distinctions between D and most of the other vitamins. Firstly, too much of it can be harmful; the resultant hypercalcaemia is an unpleasant assortment of disorders, including mental retardation. Secondly, there is not much D in food and many human diets are entirely lacking in the vitamin; it occurs in the liver oils of some fish, such as cod, tuna and halibut, and there is a little in egg yolk and milk, but large numbers of people in the world consume none of these things.

The life-saving benefit is that the sun, via its ultra-violet waves, can manufacture vitamin D by acting on the sterols in the skin. Therefore expose some skin, and the problem is solved. Either be an Eskimo, living to a great extent off fish oils, or be an African without the oils but with plenty of sun. In neither case should too much D either be consumed or manufactured. The food authorities have suggested that 400 international units (ten-millionths of a gram) are a sufficient intake per day. Many industrial communities, hidden from the sun, therefore have to add D to certain basic foods artificially, taking care not to do so to excess. A child can become ill if it consumes 20,000 units in a day, an adult after 100,000 units, and it is fortunate that there is such an ample margin of safety. However, in the tropics, and according to W. Farnsworth Loomis, an unpigmented and untanned white man can manufacture up to 800,000 units if he exposes his whole body to the sun for six hours. The result of such a dose is probably fatal.

Therefore black skin, according to the D theory, may be a disadvantage in absorbing more heat, and it may be beneficial in protecting against sunburn and skin cancer, but it is of overwhelming importance in preventing an over-abundance of vitamin D. Loomis argues that the dark skin came first in mankind's history, presumably developing steadily while the equally protective hair vanished from the body. Many anthropologists argue, for quite a different set of reasons, that mankind originated in Africa and they too are happy to accept that a black man was first upon the scene. With the migration northwards, assuming that continent to have been our birthplace, the early men would have encountered much diminished sunshine – and all the rachitic disadvantages of D deficiency. Those who were lighter-skinned would have benefited from a greater penetration of the ultra-violet waves creating a greater supply of the missing vitamin. They would have been more likely to have survived.

Certainly tropical people are dark. Certainly temperate people are white, and the relatively dark skin of the Eskimo can be explained by the abundance of fish oil obviating any need for transparency. The white races still have the ability to turn themselves a darker shade in summer, a fact upon which much of the Mediterranean economy is based, and so do the dark races, perhaps less dramatically but certainly visibly. The ability to adjust the protectiveness of the skin has therefore been both developed and maintained.

What about the so-called yellow people? Their colour is neither that of a sun-tanned white man nor that of the lighter groups of tropical people. Loomis argues that the yellows are manifesting a return to darkness after a white period in the past. Certainly those Asians that penetrated to the sunnier regions of the Americas are darker than those in the north, a change that cannot have begun much before 15,000 years ago. Moreover, like the Negroes, their babies are lighter at birth and grow darker, a method perhaps of providing more D for the growing phase. And, as with the Negroes, both their palms and soles are lighter, the two regions where sunshine does not fall.

Now to race. The basic trouble with our frequently sweeping attitude towards it is that we can be victims of what has been called typological thinking. Of course there are differences between the races. Of course there is kinky hair, and straight hair, fair hair and black hair, and all that variation of melanin in the skin. These things must have had their evolutionary value in the past, just as vitamin D – or so it would seem – was a decisive factor in a naked world living off the countryside. Similarly, there must have been other causes leading to differing proportions of blood groups between the races, to changes in stature, to a shift in hip width, to a thickening of lips, to a longer leg relative to the trunk, or a greater susceptibility towards this or that disease.

It is all interesting, but not when it is encased within typology, within alleged borders that prove to be impossible to define. There are different groupings, for there are brown people with straight hair, white people with black hair, and on and on, but no one can agree as to the number of these groupings. They merge into one another. They are not distinct. The Australoids are called archaic whites. The Ainu have as much hair as the Nordics, if not more. There are Micronesians and Polynesians and Melanesians, and no agreement exists about them. It is confusion. It is a determination to make classifications, as humans are wont to do. It is typological thinking of a high order.

The racial distinctions would be more satisfactory if there were not such tremendous overlap between them. Even J. F. Blumenbach, the early anthropologist, reluctantly admitted this point while at the same time classifying men into White, Yellow, Black, Red and Brown. In 1793 he wrote that: 'No variety of mankind exists, whether of colour, countenance or stature etc., so singular as not to be connected with others of the same kind by such an imperceptible transition, that it is very clear they are all related, or only differ from each other in degree.' We are not fish and fowl, or fur and feather. We are one species, but with considerable variation and with every reason to gain from that variation. It is the range that is important. As Ernst Mayr said: 'It is not only unfair but actually misleading to adopt a typological approach . . . It will have to be eradicated totally before we can seriously consider the future of mankind.'

Nevertheless, racial differences do exist, despite overlap and despite a frequent exaggeration of the distinguishing features. On them I cannot do better than quote from Sonia Cole's excellent *Races of Man*. There are, for example, the skeletal features:

Caucasoids. Bones of the skeleton are heavier and thicker than in the case of the two other stocks. Joints of long-bones are larger, muscular markings are more prominent. The pelvis is wider. The skull is characterized by more developed brow-ridges, large mastoid processes, a straight, or orthognathous, face, small jaws, high, narrow nasal bones and well-developed nasal spine, prominent chin.

Negroids. Slender long-bones; bones of forearm and shin long relative to bones of the upper segment of the limbs; narrow pelvis. Face marked by strong alveolar prognathism (bulging of the upper jaw in the sub-nasal region), poorly developed chin, low and broad nasal bridge, very broad nasal aperture, long and rather narrow palate and dental arch, rounded forehead, long or dolichocephalic skull, prominent occiput (back of the skull).

Mongoloids. There are no very distinctive features in the bones of the body; it is the skull which is particularly diagnostic. Brachycephalic or round skull common; malars prominent and pushed forward so that the lower rim of the orbit (eye-socket) is advanced; brow-ridges poorly developed; root of nose very flat and broad, bridge low, nasal aperture narrow; palate and dental arch short and wide; lower jaw wide with flaring hinder angles; shovel-shaped incisors (scooped out behind); flat occiput with marked torus or ridge; vault of the cranium often keeled along its length.

To try to compare this assortment of features with the limited fossil material available is virtually impossible. The origin of the groups of mankind is therefore a matter primarily for conjecture. The three major races were known to the ancient Egyptians. There were white men on cave walls 20,000 years ago, and probably the major three, the blacks, yellows and whites, had been established before the end of the Pleistocene. With the final retreat of the ice, and with a rapid change in the environment, the time was well suited to differentiation, to speedy evolution and, possibly, to the creation of many of the sub-races of today. The thirty of them, give or take a few, were all in existence, so Sonia Cole believes, before the end of the food-gathering stage in our history.

However, it is wrong to imply any constancy in our differences. Change, forwards and backwards, or parallel change, or divergent change, must all have been happening, time and time again, in the recent flourish of our adaptive radiation. Peppercorn hair is typical of Bushmen, of Pigmies (who may have been linked) and Andaman islanders (who certainly were not). Black hair is a feature of the darker races and of Mongoloid people, whether they live in hot parts or cold, but Caucasoids have fairer hair in more northerly latitudes. The Mongoloids have high cheek bones (malars) and epicanthic folds to their eyes (both thought to have been a response to Ice Age cold) but Caucasoids do not. Similar eyes among Bushmen may have been a response to the glare of the sun, but other inhabitants of the tropics are without this benefit. The Negroids and the Australoids both have jutting jaws, but are certainly unrelated. Upper Palaeolithic skulls were typically long-headed and there seems to have been a general inclination since then towards the round-headedness of brachycephaly. No one knows why.

In short, to summarize the argument, and to borrow from J. B. S. Haldane: 'Greater attention needs to be paid to human genetic diversity. There is no evidence that either individuals or races are equal in their natural endowment. Their inequalities are genetic. Any satisfactory political and economic system must be based on the recognition of human inequality.' 'From each according to his abilities,' as Karl Marx wrote, 'to each according to his needs.'

The races and groups, pinks and reds, yellows and browns, blacks and whites, Yahoos and Houyhnhnms, them and us, are indeed dissimilar. However – and this is the crucial point – the differences between the averages are usually small when compared to the differences between individuals within the same population. Some populations are taller than others, some heavier, some must be more intelligent (why not?), but they are all presumed to possess the full spectrum of human genes, whether for tallness, for heaviness or for intelligence. In other words, despite genetic differences between populations, they all have equal potential within them.

Benson E. Ginsberg and William S. Laughlin, geneticists, have estimated that any group of 30,000 people on Earth is genetically capable of re-creating every accomplishment of mankind, without genetic crossing from any other group. The number – 30,000 – is a guess based on 'the achievements of past civilizations such as the Mayan which, while isolated from outside genes, generated mathematics, astronomy, writing, architecture and the concept of zero – all from a base of primitive hunters'. Either such hunters lift themselves up from simple beginnings by their own energies, or they are adept at borrowing from others. Both ways lead to advancement.

What has made man so successful in the past 40,000 years has been his ability to acquire information, his genetic capacity for culture. 'How nice it is to *know* things,' said Bertrand Russell, and it would seem that all populations are equal in this respect. We were savages 40,000 years ago, but we possessed that capacity for culture. Recent history has merely been our expression of its inheritance.

As a way of bridging the gap between the evolution of our different populations and the recent evolution of our species as a whole there are two changes of fairly recent date that are both important and highly relevant to genetics. The first involves disease, and how it was influenced when small, isolated communities of people started to band together in populations sufficiently large for new infections to achieve a hold. The second involves migration, and how both a new environment and a greater degree of outbreeding combine to change appearances.

First, then, to disease. Genetically there is a double importance. Not only have numbers been depleted from this cause, and presumably whole populations have become extinct, but a greater ability to survive this or that disease must be with us today, whether the disease itself is still important or not. Measles is an example. This virus is extremely destructive to human life and, so far as is known, has no host other than man. It relies for its survival upon infection from human to human, but only a victim of it can excrete the virus and he does so for no more than a few days. Thereafter, providing the affliction has not killed him, he is both safe from further attacks and incapable either of harbouring the virus or of excreting it for the infection of others.

How then did it survive in the pre-Neolithic days when men lived in isolated groups? Bushmen today, or Amer-Indians still living their ancient styles of life in the Brazilian forests, are only found in small communities and at a distance from each other. The current belief is that measles could not have existed when all of humanity was living in that fashion. Frank Fenner, Australian biologist, has calculated that the disease needs a very considerable collection of people in one place if it is to co-habit, with some

3,000 cases a year being necessary to keep measles alive. In fact, measles will die out on islands (or within any truly isolated portions of land) containing fewer than 500,000 people, and has done so. It certainly died out among the Asians on the American continent, for its reintroduction there by Europeans caused immense loss of life.

Where, therefore, did the disease come from? The answer may have been the dogs that mankind was befriending (or vice versa) in the early days of settled communities. Measles could have been a corrupted form of canine distemper caused by a mutation enabling a virus formerly specific to canines to become equally specific and just as deadly to humans. At all events, the move to live in communities certainly led to measles, to smallpox and a score of other diseases. Each of them was either more destructive, or less so, according to the genetic combinations of the people they attacked. Consequently, the assortments of genes least vulnerable to each new attack were handed down in greater numbers to the future generations, and to us. The disease may have gone, or be relatively unimportant, but the inheritance is still with us, inextricably, and must be having an effect.

Just as that move towards settled communities had been a novel practice for mankind, and had opened up another corner of Pandora's box, so did the modern custom of hygiene. Poliomyelitis in earlier days was an inevitable feature of existence, along with poor sanitation and rotting food. Babies were infected with the virus fairly speedily, either dying from its virulence or surviving and becoming forever immune to that one form of viral attack. With the arrival of better hygiene, notably in the twentieth century, there was no longer such instant exposure to infection. Consequently, epidemics became possible, suddenly hitting at those who never before had been acquainted with the virus. Polio became the latest scourge. It continued to attack infants, but it could also kill the young and even the middle aged. Eventually it was put down by science and inoculation but, as with measles, with smallpox, and with other legacies from the past, the capability of withstanding the disease is still part of our confused inheritance. Once again a previous advantage is now likely to be a disadvantage.

As this point is very important it is as well to underline it yet again and finally. Almost all mutations are bound to be harmful. They are changes in the genetic material and are therefore almost certain to do harm. They have been likened to the random prodding of a finger in the back of a watch; the action may do good but it will probably do harm. Moreover, any mutation is likely to have more than one result. By adjusting, perhaps, the production of a particular enzyme there will probably be a secondary effect causing other enzyme productions to differ. Therefore, even if the mutation has one beneficial effect, it will have others and they, almost certainly, will be harmful. If the single benefit – of rendering greater immunity to a disease – is suddenly unimportant, there is no longer any virtue in the array of harmfulness that partners this resistance, but the genes continue all the same. Nothing can ever start afresh. All new genetic arrangements have to be based upon the old, upon old needs, old pressures, old patterns of existence.

Now to some more precise and more measurable manifestations of recent evolutionary change, and to stature, weight gain and the shift in menarche. One does wish that our forefathers, having been at such pains to learn to read and write, had written down rather more for us to read. When they have done so, and have measured things or people over the years, their efforts can be fascinating. There is a classic series recording height in Norway. From 1760 to 1830 there was an average adult increase of 0.4 in. (or 0.06 in. a decade). From 1830 to 1875 the increase was 0.6 in., a much faster rate (0.13 in. a decade). From 1875 to 1935 the rate was faster still, with the increase being about 1.5 in. (or a quarter of an inch each decade). A different Norwegian series, dealing with 1922 to 1962, shows that the height gain in that period was greater still, a third of an inch for each ten years.

At Harvard University, Massachusetts, familial enthusiasm for the same educational establishment has been scientifically exploited. Measurements have been made of the heights and weights of four generations of the same twelve families entering Harvard between 1870 and 1935. The original twelve students provided 24 sons, 30 grandsons and 19 great-grandsons for their alma mater, and the average birth-dates of these succeeding generations were 1858, 1888, 1918 and 1941. Increase in height was over an inch (between generations I and II), half an inch (between II and III) and nil (between III and IV). Weight increased only very slightly between I and II, and then by over 4 lb. between II and IV. This small sample therefore suggests that height has stabilized but weight has not.

Further Harvard measurements have been made and compared between 500 freshmen in the 1930s and in 1957–8. Half of each group were from private (fee-paying) schools and half were from public (state-financed) schools. Those from private establishments gained 0.4 in. in height during that time, while those from public schools shot up by 1.6 in. With weight, and during the same period, the public school freshmen expanded from 145 to 158 lb., whereas those from private schools increased from 152 to 162 lb. Therefore both height and weight (in this example) are in a state of change, although less so among the products of fee-paying schools and wealthier backgrounds. In nearby Boston a group of Italian-Americans working in a factory were also measured. Almost all of them were of Neapolitan parentage, and the height difference between those in their twenties and those in their fifties was over two inches, a staggering increase for such a short period of time.

The same trend has been occurring in other countries. For example, Charles Roberts wrote (somewhat disturbingly) in 1876 that 'the English factory child of the present day at the age of nine weighs as much as one of 10 years did in 1833 . . . Each age has gained one year in forty years'. In Glasgow it was being stated in 1950 that the 5-year-old of that year was (not just in appearance, but *was*) the Glasgow 6-year-old of 1900. The age of menarche, of a girl's first menstruation, has also been advancing – by about 3 to 4 months a decade both in Britain, several other European countries and much of America. A girl in these places today comes to sexual maturity between $2\frac{1}{2}$ and $3\frac{1}{2}$ years earlier than her Victorian counterpart.

It would seem as if the human body, whenever deprived of the right conditions for maximum growth, still somehow remembers what height (genetically speaking) it ought to be. If the conditions were not correct during childhood, the finalization of growth is remorselessly continued deep into adulthood, perhaps until 26 or so years of age. If conditions are better during the early years then growth will cease at an earlier time. Conditions were better in the 1930s for those Harvard men at private schools, and they matured to their correct height earlier than those from public schools.

However, there is still that matter of a higher adult stature when growth has finally ceased. It was Gunnar Dahlberg who first suggested that genetic dominance might be involved in today's greater heights. If there were some dominance among the (presumably many) genes affecting human stature this could affect the final height of the offspring. In other words, if one parent was tall and the other was short, the offspring would end up slightly taller than the midway point between the two parents. Parents are more likely to have a different genetic inheritance themselves if there has been outbreeding rather than inbreeding. Therefore, when a community moves out of its valley and encounters different combinations of genes, the result on the offspring will be a greater degree of mixing, of different gene arrangements from the parents. This will result, assuming dominance for stature, in slightly taller children.

Without doubt there has been more outbreeding in recent years. Even the stay-at-homes have been finding their mates from further afield, or from more than a mere field or so away, as was their custom. Migrants have been even more prone to mate with people substantially different in their origins, perhaps even from different countries, and sometimes from different continents. Hence the claim, founded on immigration and a prosperous environment, that the United States now possesses – at 5 ft 8.2 in. for men – the tallest measured human average in the world (save of course for a few bizarre communities, such as the Watutsi and Nilotic tribesmen of Africa, whose males average 5 ft 10 in. or more).

Good conditions in America must have played a part; so too genetics. Some studies have been made on communities where only the genes, and much less the environment, were important. Frederick S. Hulse, for example, discovered that the sons of parents who had come from different Swiss villages were taller by nearly an inch than the sons of parents from the same village. Similarly, to marry your cousin is, on average, to produce shorter children than if a mate had been found beyond the family. It would therefore seem, concludes Professor J. M. Tanner after surveying all the material, that the change towards outbreeding may be responsible for as much as 0.8 in. of height increase per generation.

Tallness is easy to measure, and indisputable. A character like intelligence is hard to measure, and extremely disputable. It is extremely likely that outbreeding, which has occurred so explosively in recent generations, may also have swollen our intelligence just as it has added to our stature. Fortunately for all those who prefer continual argument rather than a confrontation with facts, no one knows enough about intelligence. The IQ tests, even for their worshippers, do not have the simple accuracy of a foot-rule, and are of fairly recent origin. Even if we did have IQ measurements dating back four generations at Harvard, or into eighteenth-century Norway, or into our Neolithic past, large numbers of us would not believe them.

Nevertheless it should never be forgotten that, as well as an evolutionary past, we are also confronted by an evolutionary future. We will change. There is no stability. So long as only a portion of us contribute to the next generation, and so long as that portion is distinct in any manner, we will continue to evolve. Our life style may be unnatural but natural selection still has a firm embrace around mankind. This point will be repeated in various guises, time and time again, in the later chapters. And so it should be lest we forget it for a moment. Genetics is assuredly the past, the present and each generation's unique contribution to those who will come our way.

CHAPTER 2

Biology of reproduction

General development – Advantage of the sexual system –
Gregor Mendel – Genes – Particulate inheritance – Morgan
and *Drosophila* – Watson, Crick and DNA – Molecular
biology – The code of inheritance – Foetal growth – Foetal
damage – Teratogenic drugs – Rights of the unborn –
Newborn problems

*'He does not realise that, instead of conceiving him, his parents
might have conceived any one of a hundred thousand other
children, all unlike each other and unlike himself'*

PETER MEDAWAR

*'If you want to know, life is the principle of self-renewal, it is
constantly renewing and re-making and changing and trans-
figuring itself'*

BORIS PASTERNAK (in Doctor Zhivago)

'What do I think about sex? Oh, I think it's here to stay'

MARILYN MONROE

This is a book confronting human genetics, not reproduction itself. Nevertheless it is initially vital, or so I believe, to pick out the highlights of the
process, and certainly those of recent discovery. It is a formidable subject
– each year now there are over 20,000 papers in the scientific literature
dealing with some aspect of vertebrate reproduction – but of course it is
the means through which genetics operates. Therefore, the implications
and possibilities of genetics should be borne in mind, whatever reproductive peculiarity is being discussed.

'There is nothing,' said the lady on television, 'that modern youth
today does not know about sex or reproduction.' Modern youth is, by
this statement, in quite a different category from modern biology which
seems to encounter major unknowns at every turn. What makes the
follicle burst to release the egg in the first place? No one knows. What

actually happens to spermatozoa between their ejaculation, when they are quite incapable of fertilizing an egg, and their so-called capacitation much later when they gain this necessary ability? Even the idea of capacitation was unknown until twenty years ago, but there are countless other reproductive uncertainties. From the moment of ovulation until the fertilized egg reaches the uterus there is a gap of about eighty-five hours, and during that time the combined egg and sperm have multiplied from one to sixteen cells. The developing form is still – as in all mammals – a moving thing, dividing as it goes and in no apparent haste to attach itself to the mother who will give it birth. Eventually, after 123 to 147 hours, and when the embryo is a blastula with a lump of cells at one end and a fluid-filled cavity at the other, the mobile fragment that is a new person is ready to implant. No one knows how that implantation happens. The embryo will not attach itself unless the uterus is suitably responsive, and therefore the mother has to be in the right hormonal state. Also the uterus will not be receptive unless the embryo is satisfactory. A glass bead, for example, or a blastula model of the right size will not be welcomed by the uterus, and cannot be implanted. What chemistry is involved in this crucial attachment?

More unknowns follow. The subsequent uterine development has been called the central unsolved problem of biology today. Just how does that assortment of cells control its own differentiation, both in an orderly and consistent manner? Somehow or other, and with considerable haste, the blastula becomes a distinct and recognizably human embryo. By the end of the third month of pregnancy the growing thing, now called a foetus, has all the organs correctly arranged. The kidneys are functioning. The heart has been beating for ten weeks. Many of the reflexes are in existence. And certainly the young person, destined to be male or female from the very moment of conception, is now visibly one or the other. No one even knows how the sexes are differentiated. There is the X chromosome and there is the Y chromosome, but that is no explanation of the ability to create male and female.

There is not even agreement on the major reason for sex itself. Terrestrial life began without sex. It continued for at least two-thirds of its time here on earth without sex. Then, probably among the blue-green algae of a thousand million years ago, this form of polymorphism began with two forms contributing for the first time to the creation of the next generation.

There are two classical explanations. One is that the offspring of a sexual system can combine the merits of both its parents and might therefore be better (in the sense of being more suited for survival) than either of its parents. The second is that a sexual population can evolve more rapidly. John Maynard Smith, British biologist, has made some calculations about this second idea, and concludes that sex has no advantage if the population is less than 100 million or so individuals. Sex increases variability but by no means is this always an advantage, even at times when the environment is changing. Sex is beneficial, says Maynard Smith, only when the environment is changing radically *and* the population being affected is very large.

Suppose there is a favourable mutation that makes an animal better able to live in a changed situation. With an asexual system, where each individual replicates only itself, much time is necessary for that advantage to spread, and be selected for, among the succeeding generations. Much time is also necessary if another advantage, mutated in another individual, is then to spread and be selected for in similar fashion. With sexual reproduction both changes can spread simultaneously. This is a devastatingly important distinction.

Suppose that an asexual animal's first mutation enables it to survive better in the cold. All descendants from that one animal will benefit should the world turn colder. Suppose that the second mutation occurring in another individual produces another advantage, say a faster breeding rate; its descendants will also benefit accordingly. Unfortunately these two sets of descendants will each have only one advantage, not both. They cannot have both unless the second advantageous mutation also occurs in an individual already blessed with the first mutation. This may happen in time, but there is no guarantee and time may be in short supply. With a sexual system, because of the vital union of two individuals, their joint blessings will immediately have the opportunity to develop within a single individual. This fortunate descendant (and all *its* descendants) will therefore both survive the cold *and* breed faster.

However, the male is an extra. He eats 50% of the food. Half of the offspring he helps to create also have to be males, again eating 50% of the food. He is superfluous, save for the virtue of the sexual system. Maynard Smith feels this advantage is at its most powerful when two populations of the same species, both adapted to slightly different environments, combine to colonize a third territory at the same time. Instant mixing on a wide scale will immediately produce offspring more suited than either parent to that third environment. Such a double invasion will not be frequent, says Maynard Smith, but the effects are so important that even rare cases would justify the males' existence. In other words, even the sexual system used by mammals and every other vertebrate, many invertebrates and most plants, is neither readily explicable nor speedily comprehensible, nor universal (and some creatures only make use of it on occasion, or when times are bad). Nevertheless, as Marilyn Monroe said in her splendid statement, sex is here to stay, but even that goddess might have been weak on the theory behind its practical ramifications. She knew her art well enough but not all the reasons for it. So also did two most different people from an earlier age.

The nineteenth-century monk, living at Brno in today's Czechoslovakia, who did more than anyone else to found genetics, knew nothing of the mechanism involved. Similarly, Charles Darwin, who did more than anyone else to put evolution in its place, knew nothing about the European monk. It is extraordinary that these two men, both assembling their respective and related jig-saws, were so successful when so many key elements were missing; the evolutionist having no facts of the laws of genetics and the geneticist without any details of the actual process. The similarity ends with the effects of their respective publications. *The Origin of Species*, published in 1859, was received vehemently, either for or against. Gregor Mendel's crucial findings, published six years later (in *The Proceedings of the Brünn Society for the Study of Natural Sciences*) were greeted with some anxiety by his religious colleagues and with silence from the scientific community. By the time Darwin had plucked up further courage, and had published *The Descent of Man*, Mendel was still quite unknown. There was not even a single rebuke (or any applause) when the new Abbot of Brno quietly destroyed all Mendel's papers after his death. The former Abbot had not just been ahead of his time, but too far ahead of it for appropriate recognition.

Basically Mendel's work had proved that traits can be passed from one generation to the next both with mathematical precision and in separate packets. Before his time it had been assumed that inheritance was always a matter of blending, as if coloured water was added to plain water with the result inevitably and inextricably being coloured water of a weaker hue. Mendel foresaw genes, the differing units of inheritance (that are named, incidentally, after the Greek for race). Genes remain distinct entities. They do not blend, like that coloured water. They can produce, to continue the analogy, either plain water, or coloured water or a mixture between the two. Moreover, assuming no other genes are involved to complicate the story, they continue to create these three kinds of product in generation after generation. The packets remain distinct.

The mathematics also have a pleasing simplicity, at least in the early stages. The human blue-eye/brown-eye situation is (more or less, and forgetting about some awkward contradictions) a good and elementary example. There are genes for brown eyes, and to receive either one brown gene or two is to be brown-eyed. To receive no brown gene is to be blue-eyed. To receive one brown gene is to be brown-eyed because one brown gene has the power of creating apparently as much pigment as two brown genes. If a population of blue-eyed people (who inevitably possess no brown genes) is mated with a population of brown-eyed people (who in this instance possess two brown genes) their offspring will all be brown-eyed. Fifty per cent of these offspring will be mixed, having only one brown gene, and will therefore either pass on a brown gene or not to their offspring. In general, no one knows, without looking at his ancestors, whether his brown eyes are a mixed or a straight inheritance, whether brown-blue or brown-brown. All blue-eyed people will know that both their parents handed them no brown gene, even though their parents may both have been brown-eyed. When both parents are blue-eyed all their children (well, almost all, as there are exceptions) must also be blue-eyed.

The clarity of Mendel's vision certainly helped science, but the discovery of his work thirty-five years after it had been originally published, and the subsequent excitement, spread his name and his basic ideas far afield. It was assumed that all of inheritance was equally clear cut, with that ratio of 3 to 1, or his equally famous ratio of 9 to 3 to 3 to 1 (involving two characteristics) explaining all our genetical fortunes. So they do, in a sense, but the real situation is hideously more complex. Only a few aspects of our huge inheritance are controlled by a single pair of genes. Only a few more are controlled by two pairs. A feature like height, for example, or skin colour may be organized by twenty or so pairs. Each pair is working in a Mendelian manner, but the effect of them all working together is a bewilderment. The mathematics still have the same precision but are only for mathematicians, not the rest of us. As for a feature like intelligence, with the brain differentiated to fulfil a tremendous range of different tasks, its inheritance cannot be thought of in a simple ratio of any kind.

There are tens of thousands of paired genes within all of us, but no one knows the number. They form us. There are enough of them, and enough possible variations, to ensure that each one of us is unique. Never in the history of man has there been you before. Never in all of future history will there be you again. You are a combination that is entirely individual, and yet your genes are common to the population you live in. There is nothing unique about them, unless you are the owner of a recent mutation, and if you could somehow select and extract the right genes from the population around you it would be possible to create a replica of yourself. However, it is impossible, and even your children, because of sexual reproduction, can only spring from your loins with 50% of what you could give them were you an asexual creature able to breed partheno-genetically.

In 1900 Mendel's work was discovered virtually simultaneously by three people in three different countries: Hugo de Vries in Holland, Karl Correns in Germany and E. Tschermak in Austria. There was suddenly much talk of genes, and then of genetics when the British biologist William Bateson coined that word in the first years of this century, but what was the gene, this discrete packet of inheritance? And where was it? Attention focused on the chromosomes. No one knew about these long strands in each cell nucleus, save that they accepted staining well – hence their name of coloured bodies – and were later observed to split lengthways during the process of cell division. Perhaps the genes were arranged along them – a statement first made in 1903 – and could some-how split in two to give each daughter cell an equal share of the gene material. Thereafter this same material of inheritance would have to refurbish itself so that it could split again at the next division. And at the next. And the next.

It was T. H. Morgan in America, and *Drosophila melanogaster,* the fruit fly elevated to the laboratory from its normal environment of rubbish heaps, that jointly made the next step. The fly has four large, exceptionally clear chromosomes. Morgan, by judicious breeding, and by working with readily identifiable characteristics, was able to plot the positions along the chromosomes of the genes that controlled these features. So far so good, but still no one knew a thing about the physical nature of genes, however well they might be located along those chromosomal strands. Once again a much earlier finding was waiting in the wings to achieve its reward.

Several decades earlier, in fact when Mendel was still alive, the Swiss biochemist Friedrich Miescher had discovered the substance deoxyribo-nucleic acid, hereafter known as DNA, that formed part of each cell nucleus. Not until the 1940s did DNA receive its due, and thereby dis-place all the notions that various proteins were carrying the genetic information. By taking some DNA from dead bacteria, and injecting it into living bacteria of a different kind, the Canadian Oswald Avery and others were able to give the new bacteria some of the properties of the original bacteria from which the substance had been taken. Everyone was amazed, including Avery. Nevertheless, by experimenting in the strange world of bacteria and bacteria-eaters, the fact was confirmed: DNA was assuredly the stuff that genes are made on. From sweet-peas in Brno, to fruit flies, and then to bacteria and even to the viruses that eat them, the science of genetics may have been descending the evolutionary tree but knowledge had been ascending all the time.

By the 1950s everything was poised for the biggest leap of all. If DNA, then how? And how could that complicated molecule *both* divide itself into two identical parts, then create itself again in each new cell *and* carry the genetic information? It was Linus Pauling, chemist – but more polymath than most – who suggested a spiral form for DNA and one composed of three interwoven strands. Had he been a biologist, or so it is argued today, and had he been more immured in the doubleness of so much of living matter, he might have thought more in terms of a *pair* of twisted strands. As it happened, James Watson, aged only 24 at the time – 1953 – and Francis Crick, then 36, were the first to dream up a double arrangement for the structure of DNA that satisfied all requirements. In a 900-word article published in *Nature*, which will be read and re-read for all time, they gave details of their molecular model that not only fitted with the crystallographic picture of DNA but accorded with all the rules for fitting together the atoms of which it was known to be composed.

Here then was the gene molecule, the carrier of our inheritance, the ordainer of each and every one of us. Chemically, it was a double spiral – or helix – with long chains of sugars and phosphates acting as the banisters, and with a stairway of four different bases (adenine, thymine, cytosine and guanine) all weakly joined at the centre by atoms of hydrogen. By this strange assortment of material we are all ordained into what we are.

The molecule had not only to satisfy chemists and crystallographers but also the biologists. It had, first of all, to be able to divide into those two equal parts. Once again, Watson and Crick were not lacking in foresight (or intellectual conviction). They wrote: 'It has not escaped our notice that the specific pairing we have postulated immediately suggests a possible copying mechanism for the genetic material.' In a second letter to *Nature* they then described the mechanism. Basically the molecule was presumed to unravel itself, pulling apart at those weak central hydrogen atoms, before amassing the missing parts – those four bases, those sugars and phosphates – from the material conveniently to hand in the cell's nucleus. The double helix, after becoming single in cell division, could then make itself double once more. Arthur Kornberg, American biochemist, put the seal on this aspect of the Watson-Crick hypothesis by proving the point experimentally in 1956.

For biologists there was still the question of how DNA effected its control. How could it ensure that the right amino-acids were picked up in the right order to make the right proteins at the right time? After all, the molecule has a far more complex job to do (if only because there are so many more different parts) than a mechanic walking into a factory where all the pieces of a car are waiting to be assembled. The scale is also very different. It is important to understand that the DNA in one human cell weighs six millionths of a millionth of a gram. With the human population not far short of 4,000 million a small but disarming sum provides the information that all the DNA which caused all of us to be put together weighs a grand total of 0.024 grams (or less than a thousandth of an ounce). There are aspects of human genetics that have the same mind-numbing effect as an astronomer's talk about distance.

Nevertheless not every mind was numbed; certainly not those that first postulated and then proved the servile function of RNA, a single-stranded chemical somewhat allied in its form and certainly crucial in its assistance to DNA. There are several forms of it. First, there is the RNA that exists in a cell's ribosomes, the areas predominantly concerned with protein manufacture. Secondly, there is transfer-RNA that collects amino-acids floating about in the bulk of the cell tissue and takes them to the ribosomes for processing into proteins. Thirdly, there is messenger-RNA. This is imprinted with the right message from the DNA, and takes the message to the ribosomes where this information is then transformed, via a selection of the correct amino-acids, into the correct proteins. The system appears somewhat slick in its simplicity.

Remember, once again, that no cell has more than 0.000000000006 grams of DNA and yet ordains the proteins absolutely. To delve into those thousands of proteins is the bewildering lot of the biochemist, and not a function of this book, but even he has no idea how many kinds there are within the human frame. However, to refer again to that analogy of a mechanic making the motor car, both the DNA and its assistant RNAs are like mechanics. They are turning out a wide variety of vehicles – the proteins – each tailor-made for a specific job. It is up to the DNA merely to produce these proteins; no more and no less than that.

There is one final and tremendous aspect to the mechanism of inheritance. Just what is the code by which the DNA operates? It is easy to say that it passes information, that it instructs the RNA to pick up amino-acids in a certain order; but somehow, within its molecule, there must be a code to contain this information. There are twenty different kinds of amino-acids and the average protein has about 150 of them joined together. Therefore, for the manufacture even of one protein, there must be a considerable instruction procedure.

It is all associated with those four bases acting as the staircase on the spiralling molecule of DNA. They contain the information, and in a quite logical fashion. (It was deduction more than anything that gave the answer.) If each base is presumed to correspond with one amino-acid, then only four amino-acids can be selected – and there are those twenty to choose from. If the bases are similarly presumed to be arranged in pairs the situation improves, but still inadequately, because there are only sixteen different ways in which four items of any kind can be arranged in pairs: AA, AB, AC, AD, BA, BB, BC, BD, CA, CB, CC, CD, DA, DB, DC and DD. However, if the bases are arranged in triplets, there are sixty-four ways in which the four items can be arranged: AAA, AAB, AAC, AAD, ABA, ACA, ADA, BAA, CAA and so on. A total of sixty-four provides plenty of combinations, with more than sufficient to cover the twenty amino-acids and the forms of punctuation that must be necessary. (There surely has to be an indication when the code for some protein is beginning and when it is ending.) It was at the start of the 1960s that the triplet idea was first proposed. In 1961 the first association was made of a particular triplet of those four available bases with a particular amino-acid. By 1965 the code for every amino-acid had been learned, and so had the fact that the same code is used in the creation of all living things: man, mouse, amoeba, or buttercup – the system is universal.

There are still gaps. For example, although it is understood that there are very many yards of DNA within the chromosomes of each nucleus, and although the weight of this length can be precisely measured, it is not known how many molecules are involved in the strands. Or whether each DNA strand is continuous within each chromosome. Similarly, although genes have come a long way from the original postulate that there existed 'discrete particles of inheritance', no one yet has the slightest idea how many genes are contained within the forty-six chromosomes of man (or the different number of any other creature). Or how many atoms there might be in any single gene. Or where those few human genes are located that are known about. On which chromosome, for example, is blue eyes?

And why and how does that extra fragment of chromosome pair No. 21 have such a devastating effect when it suddenly arises to cause all the well-known signs of mongolism?

It is easy to be suffused with something tantamount to reverence when investigating any fragment of some natural circumstance that is as exceptional as the story of DNA. It is therefore as well to remember the West African in the aeroplane. It was his first flight and the engineer sitting next to him said: 'Don't you find it amazing that this machine, weighing eighty tons and carrying all of us, just sails into the air at the end of the runway?' 'Why?' said the West African, 'what was it supposed to do?' What else is DNA for, save to store information, contain the genes, pass on that inheritance and arrange for the manufacture – of a frog, or of any one of us? That is all.

The fertilized human egg, the free embryo, the implanted embryo, and then the foetus exist for a total, more or less, of 266 days. From a single cell, barely visible, to three kilos of life capable of independent existence – the thirty-eight weeks witness considerable activity. Never again will the rate of growth be so considerable, nor ever again will differentiation be at such a pace. The one cell becomes the several thousand million cells of a newborn infant, but that infant will only gain in weight thereafter some twentyfold to achieve both adulthood and an end, at last, to growth. It is unnecessary to state that the time spent in the uterus is critically important, demanding every conceivable care, but it is extraordinary to realize how little attention has been paid in the past to this foundation period of all our existences.

The body itself is less laggardly. For example, it uses the period for the dismissal of gross errors. Very many embryos (some reports say one in a hundred) are rejected without even the women who had been carrying them knowing that a conceptus had occurred. Disordered embryos and foetuses continue to be rejected at later stages, with large numbers of spontaneous abortions being of abnormal offspring. About a fifth of these have visibly abnormal chromosomes, an extreme form of disorder, and it is generally thought that most premature expulsions from the womb are malformed in some manner, even if the errors cannot be observed. As these rejections occur mainly in the first three months, and as even a twelve-week foetus measures only $2\frac{1}{4}$ in. from its crown to its rump and weighs three-quarters of an ounce, it is not unreasonable that many of the lesser abnormalities remain unseen, quite apart from the fact that spontaneous abortions frequently occur when medical observation is not at hand. At all events the body rejects, but not uniformly. Some chromosomal abnormalities are dismissed almost entirely and speedily; others survive to form part of the number of congenital abnormalities witnessed and recorded at birth.

German measles and thalidomide (alias Contergan, alias Distaval and various other guises) have probably done more between them to warn both mothers and the medical profession of the acute vulnerability of pregnancy. In 1941 the link was first spotted between the catching of this form of measles (or rubella) and the occurrence of defects, notably deafness and cataract, among the offspring of measles-infected mothers. In 1961 the soothing, nausea-reducing and cheap drug named thalidomide was first linked with a sudden spate of short-limbed babies, and a generation of 7,000 children around the world are now eloquent evidence of that unwelcome vulnerability.

By no means has the better understanding solved the problem. The United States suffered a rubella outbreak in 1964 that is thought to have caused 30,000 to have been born with abnormalities. Eyes and ears, both of which start their development on the eighteenth day after conception, are well known to be susceptible to the effects of this virus, but now there is information that the brain too may be harmed, and in a most specific manner. Normally there are about five autistic children in every 10,000 live births. For a group of 243 rubella children in New York the proportion suffering from autism (the severe form of withdrawal from interaction with other people) was equivalent to 412 cases per 10,000 – or almost 50 times the normal number. For one virus to affect ears, eyes and some portion of the brain, and for a drug, chemically simple, to have the most specific effect of drastically shortening arms or legs, is to be made immediately aware that the foetus must be damaged with similar severity by a wide range of other products, diseases, foods and drugs.

The list of drugs for which there is satisfactory evidence of a possible harmful effect on the foetus is formidably long. (One taken from a recent issue of *Medical News-Tribune* is in Appendix 1.) Many of the names may seem to be excessively polysyllabic and unfamiliar, but within their syllables are ingredients well known to all of us. Salicylates, for example, are the principal ingredients of aspirins. Barbiturates are better known, but so they should be considering that the western world consumes a ton of them every day for every million people, Mycins, cyclines and penicillins are all antibiotics and they too figure on the list. It is therefore very easy to wonder again about the foetus, warm and nourished within its amniotic pool, but bombarded pharmaceutically from the very start of a nine-month siege.

As an article published (in 1972) in the *British Medical Journal* modestly put it: 'There is growing interest in associations between events in pregnancy and the subsequent health of the child.' Or, as another article phrased it: 'A pregnant woman should take no drugs, and even fewer during the first three months.' In one American survey, the average number of drugs taken by a group of pregnant women was 8.7 each between their third and ninth weeks, the most vulnerable time of all. An Edinburgh survey of 458 mothers and abnormal children emphasized that they had been taking well above the average number of such simple drugs as aspirins, antacids, barbiturates and amphetamines. Perhaps mothers more likely to produce abnormal children are also more likely to take more drugs, but the finding seems to be yet another warning. Little enough is known about the effects of any drugs even upon an adult's central nervous system, and nothing is known of their effects upon the foetal cns. Hence the growing feeling of alarm. 'Obviously this should not lead to therapeutic nihilism,' the British gynaecologist Philip Rhodes has written, 'but to a constant awareness that the use of drugs in pregnant women might damage the foetal brain.'

There are also drugs that are a direct problem in themselves, such as heroin (low birthweight for the foetus, fits, vomiting, respiratory distress), lsd (chromosome breaks, stunting of growth) and, needless to say, the most favoured drug of all, the inhalant that has swept the world. Herewith a *British Medical Journal* leading article (of 1973): 'No reasonable doubt now remains that smoking in pregnancy has adverse effects on the developing foetus. The effects range from retardation of foetal growth and prematurity to an increased risk of perinatal death from all causes . . . It is the smoking itself rather than the type of woman who smokes that is responsible for these effects.' A report from Sheffield added that 'two of every ten unsuccessful pregnancies in women who smoke regularly would have been successful if the mother had not been a regular smoker'. To conclude the diatribe, herewith the *BMJ* again: 'It seems the time has come when women should be told frankly that if they smoke they not only put their own lives in jeopardy but, if they continue to do so in pregnancy, expose also their unborn infants to an unnecessary risk.'

Quite suddenly, prompted no doubt by an increased awareness concerning foetal hazards, there is concern for the rights of the foetus itself. Can such an offspring sue anyone, such as the doctor or even its mother? It has rights the moment it is born. Should it not have rights before it encounters that moveable feast of birth, that switch from being a non-breathing creature fed through its umbilicus to a breathing creature taking sustenance through its oesophagus?

When (in 1973) the Supreme Court of the United States decided that a woman's freedom was inalienably linked with her right to have an abortion during the first six months of any pregnancy, the same edict contained an unprecedented declaration concerning her unborn offspring. It was the first to confer such rights provided the foetus was older than six months from the time of conception and even if it was never subsequently born. The statement therefore raised a crucial issue: would it be possible to found a criminal charge against someone who harmed a foetus of between six and nine months?

In no other country is there such a ruling, although the Supreme Court's decision still has to suffer the experience of its handling in lesser courts, and other countries are waiting to see what might happen in America. Certainly in Britain there is no such law, and not much in the way of legal precedent. There was, however, a case at Liverpool in 1939. Negligence caused a ladder to fall on a pregnant woman. The accident caused the child to be born on the following day and it lived for only one more day. The parents, as administrators of the dead child's estate, brought a claim for damages for loss of life expectation. They received a settlement of £100 from the defendant who had caused the ladder to fall. (Perhaps that figure should, in parentheses, be compared with a settlement of £321,000 awarded by one American court to a thalidomide child.)

In a legal textbook it is written that those dealing with the right of a plaintiff to recover for foetal injuries are 'at once Janus-faced and marred by obsolescence'. In fact, Janus is having to look more ways than two on this subject; events are happening in every direction. The medical journals are steadily reporting both safer and later dates for abortion. The obstetric wards are equally steadily reporting their ability to keep premature babies alive that have been born at earlier and earlier times. There is a move, in Britain at least, that a gestation period of only twenty weeks ($4\frac{1}{2}$ months) 'should be regarded as prima facie proof of viability at the present time'. As yet no baby born younger than twenty-five weeks has survived but, as the British paediatrician Neville Butler said: 'Who is to know that within the next two decades viability will not be down to 20 weeks. When planning ahead it is good to allow some latitude.'

With the courts, and plaintiffs and defendants, and hard cash and publicity all involved, the foetus is joining the society of which it is a part earlier than ever before. In the past it appeared, dramatically, for good or ill on that day when its lungs were filled with air for the very first time. It is now becoming a person with some rights at six months, or at twenty weeks, or even at conception, or – perhaps – even before then. If two parents, counselled of a genetic malformation lying in wait for a conception, go ahead and produce what had been predicted, what then of rights, of negligence and of a malformed plaintiff suing those who chose to make him? The foetus has not been much of a person until now, but his day has assuredly begun.

With the arrival of birth the die is not only cast but seen to be cast. The infant is there, crying, protesting, and either with or without its correct apportionment of faculties. Parents may abhor the very idea of eugenics, that anyone should dare to suggest ways and means either of improving an offspring or of discouraging defects, but within minutes they will have asked about fingers and toes and will (probably) have been given assurances. Grandparents, equally antagonistic to eugenist notions, will first ask about sex and weight and then expect similar comforting remarks about the normality of their descendant. Inheritance is roulette with almost everyone unwilling to be informed how the table might be loaded. Should misfortune result the parents will, far more often than not, accept their fate with humanity and welcome the child with much compassion, with considerable love. In the gamble of conception they have lost by any normal yardstick, but they can only see the event most positively. Their strength is astonishing, as they extract every ounce of satisfaction from the extraordinary gaiety of a mongol, from the will and determination and spirit that seem part of every handicap.

In roulette the player himself loses. With inheritance the players may see their lot as gain, but what of the child? What purpose is there, as with Tay-Sachs disease, of living for at least six months but of dying within three years? What is the gain of the pain of haemophilia? The new abortion laws were founded largely upon concern for the misery of an unwanted child. There is far less concern about unwanted disabilities, or the prospect of a paralysed existence. We abort tons of foetuses and then lavish every medical aid upon babies merely because they have been born. Consequently the feeling is growing that preservation of life at all costs, whatever the nature of that life, is not the only criterion. If it is humane to abort, is it not also humane to reject?

Medical hysteria is the name given to the syndrome under which medical men carry out procedures on patients primarily for the sake of doing something. Few doctors, says Eliot Slater, are immune from its ravages and 'we are practically all of us suffering from a hysterical blindness in which we refuse to see the facts as they are. Just because something can be done there is a feeling that it should be done'. Medical science can, for example, keep even hideous monsters alive for several weeks. It frequently does so, but such products have no future and, save for those who see reason in everything, they have no purpose. Is the veterinary surgeon less or more humane for hitting similar deformities on the head the moment he encounters them? 'We need to prevent the investment of human care and the sacrifice of human happiness in a lost cause,' adds Slater. 'We should cut out this waste at the earliest possible stage – prevent the abnormal gamete from becoming a zygote, prevent the abnormal embryo from being born, prevent the abnormal neonate from becoming a child . . . No good case can be made for taking the moment of birth as an ultimate time limit, after which no further pruning is to be allowed.'

If a newborn child dies it is a tragedy. However, it is a loss that the parents can probably remedy in a year, and a severely handicapped child can be replaced by a healthy one. 'Medicine,' says the geneticist J. M. Thoday, 'has been interested in length of life rather than quality.' Theodosius Dobzhansky, putting the other point of view as well, has written: 'If we enable the weak and the deformed to live and to propagate their kind, we face the prospect of a genetic twilight. But if we let them die or suffer, when we can save or help them, we face the certainty of a moral twilight.' It is questionable whether many of today's actions, prompted by a kind of Hippocratic inertia, either save or help those most concerned, namely the parents and the offspring they have produced.

Extremely premature babies are an example. For some reason or reasons the uterus has chosen to expel them long before normal viability has been reached. Then, for no particular reason save the ancient momentum of wishing that all life, particularly young life, should be prolonged, the half-finished infant is given intensive care. One problem is to give the right amount of oxygen: too little, and the brain can be damaged; too much, and there is retrolental fibroplasia, a form of blindness. 'If a choice must be made, I tend to give concern for the brain first,' said a doctor. The same man, reporting on his Canadian hospital, which had paid well-above-average attention to the problem of perinatal death, said that the mortality rate for babies born weighing between 1,000 and 1,500 grams (2 lb. 2 oz. to 3 lb. 5 oz. – about a third of normal) had dropped to 46%. In the years 1968–69 the rate had been 82%, despite great care and expense, but the new percentage of roughly 50–50 was 'the best we have been able to do with this group'.

Premature babies (less than 1,500 grams) are also more likely to be abnormal. Even those that survive, presumably those more fortunately endowed, have many more defects than full-term children. Facts taken from Canada, England and elsewhere have shown that 85% of the survivors are considered as normal when examined one year after their traumatic start to life and about 8% are definitely abnormal. The ratio is at least four times, possibly eight times, worse than average even among the survivors. Any argument about the need for swift replacement of babies, rather than the protected care of the least fortunate, carries even greater weight with prematures; they were carried in the womb less long, and less time (and longing and hope) has therefore been committed to them.

This reproductive section now ends with two quotations, one from each side of the fence that has been implicit in these recent pages. It is sometimes easier to come to a conclusion when hearing the opposition standpoint rather than a statement nearer to one's own tentative beliefs, but both can assist in their differing fashion. Here then are the two opinions, the first from a medical researcher, the second from a consultant surgeon.

'The grossly damaged neonatum, judged by its effects on the family in which it is reared, is a malignant parasite. The factual evidence on this is unambiguous and convincing. Families that rear a spastic or a mongol or a phenylketonuric or any severely mentally-handicapped child suffer grave social and psychological stresses; the mothers suffer in physical and mental health; the families tend to become isolated, and within them there is resentment, shame, guilt, over-devotion to the child, quarrelling and sometimes breakdown of the marriage.'

Secondly (and conversely):

'We know that children who are severely disabled by congenital heart disease will die before they are three if we do nothing. Due to new methods of co-operative technique we can often bring about a very substantial normalization of cardiac function.'

On the one hand a plea for grasping the nettle more firmly than ever before; on the other a hope for 'substantial normalization'. It is up to us to decide which course is preferable, both now and in the immediate future.

CHAPTER 3

Mating and marriage

Custom – Interbreeding and inbreeding – Incest and
consanguinity – Leviticus – Marriage laws (Mohammedans,
Christians, Jews) – Forbidden pairings – Miscegenation and
Negroes – Random mating – Assortative mating – Sexual
selection

'Marry one who lives near you'

HESIOD

*Italians used to find their mate at a distance, on average, of
600 yards. With the invention of the bicycle this average leapt
up – to 1,600 yards*

*'Thou shalt not let thy cattle gender with a diverse kind: thou
shalt not sow thy field with mingled seed: neither shall a garment
of linen and woollen come upon thee'*

LEVITICUS XIX. 19

Marriage, from the viewpoint of human genetics, is of no consequence
unless it is synonymous with mating. There can be mating without
marriage, and marriage without mating, but there can be no next generation, of course, without conception. Nevertheless there are facts about
marriage itself that have considerable genetical relevance. 'If anyone
thinks that religion, wealth or colour are matters to be taken into account
when deciding if a certain marriage is suitable, let him dare to suggest that
the genetic welfare of human beings should not be given equal weight,'
said Peter Medawar.

Most relevant are the various customs that prevent incestuous relationships, the results of such inbreeding, the degree of affiliation, the likelihood of shared genes between the two partners, and the levels of assortative as opposed to random mating. Every society has its laws, and every
individual feels that he or she has complete freedom of choice beyond these

general impediments; but the laws change from society to society, they change with time, and individuals are not quite so randomly behaved in their choice of a mate as they might suppose. It is therefore pertinent to examine marriage laws, incest taboos, sexual attraction in general, and the predilection for like to marry like.

Firstly, a word about humanity in general. *Homo sapiens* is only one species; any individual can mate successfully with any other individual, however different or distant the two of them may be. Within the animal world this is not always so: there are some species that can only mate with their neighbours. Assume, for example, a world-wide distribution of one species; the animals of this kind living in Europe can only mate with each other or those near by in the Middle East, but these can also mate with the Indians, and these Indians can also mate with Australians but not with the Europeans. In such animal species the Europeans at one end of the scale are not only incapable of fertilizing but have no urge to fertilize the distant members of the same species. With mankind, on the other hand, there is neither an inability nor a lack of enthusiasm. Colonial man has vented his desire, happily and successfully, upon whatever sort of woman he has encountered. Social results may have been abysmal, but physical results have always been as harmonious as with his own kind. European man and Australian woman, for instance, although distant in many respects, create offspring satisfactory in every degree.

There can be unsatisfactory crosses, just as there are with dogs. Many groupings of mankind are physically small, others are large, and their pairings can be unfortunate. For example, the burgeoning child sired by a large man can be too large for the small woman's uterus. A caesarian is then necessary if the child is to live. What is important is that this unfortunate union is no more disastrous than between the extremely big and the extremely small of any group, distant or close at hand. The conflicting stature is at fault and not some conflict of kind.

There are several reasons for the singleness of humanity. Firstly, the distribution of modern man, although wide-ranging, is a fairly recent event. It is thought that the Australian aborigines, for instance, only reached their new continent about 8,000 years ago, or 320 generations before now. The Amer-Indians are similarly thought to have arrived in their new world about 20,000 years ago, or 800 generations before today. Such periods of time are short in the history of evolution, and exceedingly abrupt for the necessary number of mutations to occur, to be selected for, and to be maintained in isolation so that two species can result. By contrast, the earliest horse, *Eohippus*, lived in the Eocene. It was a creature markedly different from today's equines, for it had five toes on each foot, a rounded back and a modest bulk; but the Eocene was fifty million years ago and all the horses from *Eohippus* onwards could have produced a few millions of generations during those fifty million years.

Secondly, unlike the horse, man has exhibited a strong enthusiasm for trading his women, for stealing them should barter encounter resistance, and certainly for sowing his specific brand of seed at any opportunity. In other words, he has always been ready to break down the isolation necessary for two species to develop. Geographical isolation can cause two such kinds, but man has always been ready to paddle across to the next island. Behavioural isolation can cause two species, if one group chooses to live in the forest while the other stays on the plains, but man has been willing to adopt or change a life-style should any reason exist for doing so. At all times he has maintained what the geneticists call a gene-flow; hence his genes can still flow, and do so, all round the world, exceedingly effectively.

It is also thought that speech had much to do with the cross-mating. Carleton Coon, the American anthropologist, has argued that groups of normal creatures are kept apart genetically, even if occupying the same territory, partly by having a set of signals with which they can communicate with their own kind. 'There is little desire,' he says, 'for them to interbreed. But man communicates by speech.' He can talk with his neighbours. He can learn several languages without difficulty, provided he starts early enough. He can teach others his language. He can steal children, and therefore their genes, from other areas, and then raise them as his own. He can constantly erode the genetic barriers, and interbreed, and hybridize yet further his hybrid kind.

Had the crossing of the Bering Strait occurred earlier, for example, mankind might have developed into two species. The citizens of the New and Old Worlds would then have been entirely distinct, much like the horse and the donkey of today. Perhaps they would have been able to create infertile mules, perhaps not. Such speciation might actually have happened in the past, and perhaps it did with the famous Neanderthals. They were distinctive people, big-boned, big-brained (larger than us) and with a fondness for burying their dead. Suddenly they disappeared, and at the same time there arrived Cro-magnon, less bony but more artistic, less solidly built and undoubtedly our ancestor. Did there exist two species, with one surviving at the expense of the other, or was there an intermingling and cross-fertilization with the heavy Neanderthal features somehow vanishing in the process? One longs to know.

Society, in general, has permitted mating over a wide spectrum, but has in virtually every grouping forbidden it between close relations. Contrarily, when the relationship is considered too distant some societies have either disallowed mating or merely frowned upon it. 'Marry one of us; there are lots of nice girls to choose from', is a message that must have been said, repeatedly, in the past. Inbreeding may be taboo but outbreeding is generally within understood limits and not too far out. As Professor C. D. Darlington put it:

> 'Men and women who marry require one another to be not merely, like animals, within geographical range and perhaps of a similar age. They also require one another (as a rule) to speak the same language, even the same dialect and with the same accent. They require, as a rule, to be of the same religious sect, even though the sect may be small, and to be of similar social and economic status, even though the status may be rare and exalted. They are also very likely to marry within a community of a similar craft or trade or profession. The tests they apply to one another all have a large genetic component . . .'

This prejudice of wishing to marry a similar neighbour is traditional, with like preferring to mate with like, and then begetting like to complete the uniformity.

The prejudice against too much similarity, particularly of blood, also has a long history. The taboo is deeply embedded within us, and the mere mention of incest caused Emile Zola to say 'it was so stupendously vile that I cannot decently contemplate it'. In England about 300 men a year not only contemplate it but their actions reach the ears of the police (and therefore figure in the statistics). One suspects, when sleeping arrangements were more communal than today, and when a father could suddenly discover that his daughter on one side was warmly and firmly reminiscent of his young wife, now snoring on the other, that consanguinal connivance of many kinds was commoner than today. One can only suspect, perhaps uncharitably, but one knows that cousin marriages were much more frequent in the past and the general level of inbreeding was closer than now.

Strangely, incest in England was not a crime at common law or at all until 1908, when the Punishment of Incest Act was passed. In Scotland a similar Act had been passed in 1567. (One immediately wonders if this was prompted by the infamous Sawney Bean, a drop-out of those days who retired to a cave near Ballantrae in Scotland, found a wife, murdered passers-by for food and clothing and, somewhat of necessity, permitted his children to mate with each other. Eventual discovery caused a judicial slaughter of all sixty-four of them and possibly the passing of that Act.)

It is the eighteenth chapter of Leviticus that defines for Christians the forbidden relationships; but the meaning is not easy to unravel. For example: 'The nakedness of thy sister, the daughter of thy father, or daughter of thy mother, whether she be born at home, or born abroad, even their nakedness thou shalt not uncover.' There is scope, one feels, for misinterpretation; so too with 'Neither shalt thou take a wife to her sister, to vex her, to uncover her nakedness beside the other in her life time.' The *New English Bible* (first published in 1961) helps to clarify both statements. First, 'You shall not have intercourse with your sister, your father's daughter, or your mother's daughter, whether brought up in the family or another home; you shall not bring shame upon them.' Second, 'You shall not take a woman who is your wife's sister to make her a rival-wife, and to have intercourse with her during her sister's lifetime.'

Arguably there is still opportunity for disagreement. That phrase about 'home or abroad' is interpreted in the *Catholic Encyclopedia* as 'equivalent to in or out of wedlock'. And does not that second quotation about the wife's sister refer to bigamy, or at least extra-curricular activity of some kind, rather than incest? Moreover how odd it is, at least from the genetic point of view, that a man's father's brother's wife is forbidden to him, namely his aunt by marriage, while he is permitted to uncover the nakedness of his niece. The aunt is a relatively distant relation, sharing no genetic material with him, but a man and his niece are far closer, having 25% of their genes in common. Many Jews and other close followers of Leviticus have in consequence married their nieces but have had to keep clear of any aunt, or brother's wife, or son's wife, all of whom are genetically quite distinct from them.

The strictures against incest, whether logical or faulty, are widespread, and therefore it is tempting to seek a common reason. Medawar says: 'It is difficult to see why the prohibition should have arisen to some extent independently in different cultures unless it grew out of the observation that abnormalities are more common in the children of marriages between close relatives than in children generally.' Even so, every society has also been capable of breaking the rules, particularly in its higher echelons where, one might have thought, the creation of abnormalities was yet more serious. (The chapter dealing with the Hapsburgs and royalty in general investigates this problem.) Maybe the ban on incest was for social rather than genetic convenience. If a man marries his daughter there can be immediate disruption in the home whereas the genetic disadvantage is not inevitable and will take much time to be manifest. If a man today were to sleep with his sister, his daughter or his aunt it is easy to imagine instant repercussion disabusing him of any follow-up. Is that explosive resentment a deep-seated anxiety for any progeny of the union, or mere dislike or jealousy or envy of a bit of poaching nearer to the hearth than is tolerable?

If genetics were the reason for prohibition one might expect that the marriage bans would apply only to close blood relatives rather than relatives by marriage. All primitive communities, co-habiting with domestic animals and aware of many breeding principles, must have known about degrees of kinship, about the difference between blood relations and the other kind. They must have known which partnership was a form of inbreeding, which was not and, if Medawar is right, which would have led to a greater proportion of abnormalities. On reflection it is easier to suspect that social peace was the prime motivator. Most children would die anyway, with or without a greater share of malformations, but peace now and within the camp was forever crucial.

The Koran is clearer than Leviticus. Also its laws are more suitable for writing in table form, set beside the possible reason for them, whether genetic or not and, if genetic, to what extent there are shared genes. For closest relationships the reason might be genetic *and* social, but the purpose of this table is to demonstrate the presence or absence of a genetic link.

In Surah IV, verse 23, and on the subject of marriage laws, it is written: 'Forbidden to you are —

your mothers	(genetic link: half genes in common)
and your daughters	(genetic link: half genes in common)
and your sisters	(genetic link: half genes in common)
and your father's sisters	(genetic link: quarter in common)
and your brother's daughters	(genetic link: quarter in common)
and your sister's daughters	(genetic link: quarter in common)
and your foster mothers	(no link)
and your foster sisters	(no link)

and your mothers-in-law	(no link)
and your step-daughters who are under your protection (born) of your women unto whom ye have gone in	(no link, as they are her daughters, not yours)
but if ye have not gone in unto them, then it is no sin for you (to marry their daughters)	(no link, as they are still not *your* daughters)
and the wives of your sons who (spring) from your loins	(no link – they are your daughters-in-law)
And (it is forbidden unto you) that ye should have two sisters together, except what hath happened (of that nature) in the past.'	(no link irrespective of whatever happened in the past)

Unlike Leviticus nieces are forbidden; they have one quarter of their genes in common with you. However, in general, Leviticus is more restrictive. This Surah verse does not mention your grand-daughters; they too have one quarter of their genes in common with you and are therefore, genetically, as close as nieces or blood-relation aunts.

The prohibitions associated with the religions of Europe and the Mediterranean are made more complex – and less ready to yield up the original motives – by the fact that change has been so frequent. For example, in the early days of Hebrew history, the laws were more lax. Abraham married his half-sister, Jacob married two sisters and, prior to the Revelation at Sinai, only a few kinds of partner were prohibited: one's mother, one's father's wife, all married women and one's own mother's daughters. Subsequently, the forbidden pairings of Leviticus (verses 6 to 18) were transformed into Jewish law. All so-called primary prohibited degrees of marriage (near blood relations and a wife's near blood relations) were extended to create 'secondary prohibited marriages'. This was a stepping-up process: the mother was forbidden, therefore so were the grandmother and the great-grandmother. At the other end of the scale this meant that the great-grandchildren of a brother and a sister, i.e. third cousins, were also forbidden to each other. Even spiritual kinship was said to exist; a godfather, for example, was forbidden to marry the child at whose baptism he had been sponsor.

At the same time the Jewish sectaries and various Christian churches and Roman law were behaving independently, extending or lessening the various prohibitions, giving dispensation when it was warranted – a feature of Catholicism rather than Judaism – and either dissolving marriages or stating that they had never taken place, deciding that they had been no more than 'an execrable action' or an 'enormity' if they had transgressed the law. For the first time children were deemed either legitimate or not, according to the degree of their parent's misdemeanour.

As a further and most relevant difficulty in the counting of degrees of consanguinity not everyone, and not even everyone in the same Church, used the same system. Genetically, the two kinds were very different. The Roman system computed degrees of relationship by adding *both* lines of descent from, or ascent to, a common stock. Normal brothers, by this procedure, were said to be related to the second degree of the collateral line. The German system counted only one line of relationship to a common ancestor; therefore normal brothers, by this method, were related in the first degree of the collateral line. Consequently, when the Council of Toledo (AD 531) is forbidding marriages to the seventh degree, or when Pope Gregory is writing to St Augustine (in AD 601) that marriages are acceptable to the third or fourth degree, or when St Peter Damian (eleventh century) is putting them back to after the seventh degree, and so on, back and forth, it is easy to be at a loss as to the actual ruling in terms of cousins and aunts.

However, one is most definitely not at a loss in realizing that genetical thinking was not at the back of it. It was ecclesiastical controversy, appeasement (for newly converted peoples), greater strictures (when liberality raised objections) or mere conflicts of personality. It was not a weighing in the balance of possible harm to future offspring. It was not an application of animal facts about inbreeding to the breeding of mankind. It was a set of laws that had everything to do with the strictures of the Church, and nothing with those of genetics.

The principal result of this see-saw of earlier edicts is that countries in Europe today have ended up with different interpretations of the Church's ruling. Henry VIII of England, as part of his all-embracing reformation, much reduced the impediment of consanguinity. Edward VI then repealed, in part, Henry's actions; Mary repealed the lot, and Elizabeth I then had to restore all her father's modifications. Everything in her country remained much the same until 1949 when ten forbidden relationships (for example a man with his deceased wife's father's sister, or a woman with her deceased husband's brother's son) were suddenly permitted. In 1960 there followed a further minor liberalization, and there are now twenty forbidden pairings, ten of which have common genes and ten of which do not. This equality of 50% is plainly unhelpful about the origin of the mating laws, and whether genetics or social conflict was the prime consideration.

Other countries have been left with other legacies. In France, the close pairings of uncle/niece and nephew/aunt are forbidden, but the head of the state can grant a dispensation. In Germany, consanguinity is a bar only in the direct line, and also between brothers and sisters. Everywhere, even in the most conservative Catholic countries, there has been a relaxation from the medieval days when the authorities seemed to be vying with each other for increased stricture. The states of the United States reflect the varying European opinions, with a few of them permitting uncle/niece and aunt/nephew marriages (there is greater prejudice against the latter) and a minority forbidding first-cousin marriages. Once again the facts of genetics do not appear to have been invoked as the guiding principle.

The crime of incest, originally punishable only by the Church, relates to no more than a few of the prohibited kinships. That 1908 English law, enacted when thirty different relationships were forbidden in marriage, only specified intercourse between four of them as incest; namely a man with his grand-daughter, daughter, sister or mother, or a woman with her four equivalent relations. At least there is sexual equality in that the maximum punishment is seven years for either men or women found guilty. Half-brothers and half-sisters are counted as brothers and sisters, but step-parents and step-children are not included. Incest is therefore counted as a crime only if blood is involved.

Adoption has provided a further problem. In Britain, for example, and in the Adoption Act of 1958, the matter of possible incest is not mentioned. Therefore, although adopted children have full rights as children, it would seem following a test case in Scotland in 1970 that intercourse between a man and his adopted sister is not an offence. Nor is it with a couple both adopted by the same parents, provided that the 'real' parents of the adopted pair are not the same; the matter of blood is held to be relevant. Perhaps the science of genetics is at last having, in a most roundabout fashion, an effect upon people's thinking. Blood and genes are finally acquiring the fresh approach that they deserve.

As one further point concerning forbidden pairings there is the bewildering history of the Negro in the United States. There exists the primary genetical confusion that an American is said to be a Negro if (according to the *Negro Handbook* compiled by the editors of *Ebony*) he has enough African blood to be identified on sight as being not white or, however white he may appear to be, identifies himself as a Negro. A man may be one-sixteenth Negro and yet consider himself and be considered not-white despite the bulk of his inheritance. A similar person, due to the genetic lottery of inheritance, may appear whiter and therefore be classified as white. Consequently all facts about Negroes and whites emanating from the United States should be viewed with this peculiarity of definition very much in mind. Some whites may in fact be more Negro than some Negroes and it has been estimated (according to the *Encyclopedia Americana*) that 10 million current whites possess Negro genes. If geneticists had to classify the American nation either as Negro or white, with no intermediate classification (and forgetting about other ethnic groups), they would presumably call anyone a Negro who was more than eight-sixteenths Negro and anyone a white who was less so; but geneticists are not in charge of the situation.

Under the current definitions there are now more Negroes living in the United States than ever before, although the proportion has fallen. In 1790 about 20% of the total population were said to be Negro; the figure is now about 10%. No one knows how many of the Negroes are wholly Negro in the sense that no white genes are in their ancestry, but this proportion must be quite small. The mixing has been considerable, despite extensive legislation aimed at preventing it. Maryland was the first state to enact a law (in 1664) prohibiting marriage between white women and Negroes. Virginia followed suit in 1691, Massachusetts declared all intermarriage between whites and Negroes illegal in 1705 and the rest of the eighteenth century witnessed a steady succession of enactments among the states forbidding, in one way or another and with differing forms of punishment, Negro/white relationships.

In 1840 the restrictive tide began to turn. Massachusetts was the first to repeal its law. Indiana vacillated: in 1840 a law forbidding marriage between whites and persons with one-eighth or more Negro blood was passed with strict penalties for violations. In 1841 the penalties were repealed but the prohibitions remained. In 1842 the penalties were re-enacted only to be repealed again ten years later. Following the civil war the southern states tended to entrench their restrictions – South Carolina re-enacted a law forbidding intermarriages in 1879, Mississippi did likewise the following year – while the other states tended to relax their prohibitions. Rhode Island repealed a law forbidding intermarriage in 1881, Maine and Michigan in 1883, and Ohio four years later.

By 1964 there were nineteen states still with statutes prohibiting marriage between whites and Negroes. These were: Alabama, Arkansas, Delaware, Florida, Georgia, Indiana, Kentucky, Louisiana, Maryland, Mississippi, Missouri, North Carolina, Oklahoma, South Carolina, Tennessee, Texas, Virginia, West Virginia and Wyoming. By then all other states had repealed their laws against mixed marriages, save for those nine states that had never had such laws on their books. These nine were: Alaska, Connecticut, Hawaii, Minnesota, New Hampshire, New Jersey, New York, Vermont and Wisconsin.

It was on 12 June, 1967 that the Supreme Court struck down Virginia's 'Racial Integrity Law' banning marriages between whites and Negroes. The Court's ruling dealt specifically with the case of a white man and his part-Negro, part-Indian wife who had been married in Washington, DC, in 1958, and who had then been prosecuted after returning to their home country in Virginia. The Supreme Court's ruling was considered afterwards to be sufficiently broad to void the anti-miscegenation laws of other states. Later that same year, for example, the first legal Negro-white marriage in Tennessee took place at Nashville between a 29-year-old Negro man and a 34-year-old white woman. At long last the United States had accepted that ethnic groupings should not provide a barrier to the legality of marriage.

Now to mating, insofar as it is permitted by law and not prevented by reason of distance, social custom or prejudice. Now, in short, to random mating, to assortative mating and to sexual selection. They are all involved in every society. They are involved when a man walks, for example, into that well-known selection ground of a dance hall. For one thing the dance hall is in his area; it is not the other dance hall down the road. Secondly, he then looks over the assembled assortment, while the assortment is equally looking over him, and he either chooses, or is chosen by, a particular companion. Is the choice random? Or does the couple have to be of similar size leading to assortative mating? And is sexual selection the key requirement for both of them?

In general terms, and as a species, we experience random mating, but within limits. Eskimos never marry Bushmen because they never meet. Northern Europeans hardly ever marry Africans and Asians because they do not encounter each other sufficiently often. Parisians do not marry Berliners or Londoners as much as they marry other Frenchmen, and they are much more likely to marry Parisians than any provincial kind. A man or a woman may think his or her choice of mate was culled from a wide spectrum, but in all probability the couple were both living near each other when they met and decided to marry. All such partnerships, even if only from across the street, still come under the heading of random mating. Within the limited choice available (how many girls were there at the dance hall? how many has a man even spoken to before he makes his choice?) the selection is at random. This, by and large, is our method.

Nevertheless, there is also assortative mating, either the preference of like with like or a rejection of similarity with a subsequent pairing of two opposites. Either people of large stature marry each other more frequently than mere chance would dictate, which is certainly the case in many societies that have been examined, or there is rejection of a similar trait in the possible partner. It would appear that red-heads experience this repulsion because there are fewer marriages between them than chance alone would dictate. We may think we are choosing at random, and select girls because they have long legs or short hair, but we possibly see in their faces a reflection of our own, a sort of matrimonial narcissism making them most desirable. We may feel more comfortable in the company of someone of similar stature, or of similar intelligence, or of similar reaction to the environment. 'I don't know why, but I can only sleep with Communists,' said a friend.

Anyway, tests have been conducted concerning homogamy, or the similarity of physical characteristics between married partners. For example, weight, stature, chest circumference, sitting height, head circumference, cephalic index, facial index, nasal index, hair colour and eye colour have all been examined. It would seem, although the correlations are weak, that there is some assortative mating with all of these kinds of physical feature. Not all the correlations are equal: hair colour and eye colour are more likely to pair off uniformly than head, face or nose shapes.

The correlation for stature is the strongest of all, but unfortunately the fact is not so simple as it might suggest. Firstly, there is a strong impulse for people of the same socioeconomic group to marry each other, and it is a fact that the more prosperous groups are taller. Therefore, to what extent is a man marrying a girl because her stature is appealing to him or because he only meets girls from his own group and they are, on average, bigger than girls in other groups? Secondly, it is a fact that people are getting larger. Therefore, as men and women tend to marry someone of roughly their own age, the stature correlation may arise (in part) because people marry within the same generation. Fifty years ago people, on average, were shorter; therefore shorter people were pairing together. These further facts somewhat erode the basic principle that people of similar stature tend to marry each other, but the principle still stands, although weaker than before.

With deafness the association is strong. It is alleged that marriages between deaf and normal people tend to be unsuccessful, and it is a fact that marriages between two deaf people are not only common but also tend to be successful. One might think, as most deafness is hereditary, that these partnerships would inevitably lead to a greater degree of deafness in the next generation, but there are thought to be about thirty-five genes that can cause deafness. Only marriages between deaf individuals carrying the same genetic defect will produce deaf children and this extra correlation, due to the large number of genes involved, is not certain.

There is also a strong correlation for intelligence. Or rather there is a strong correlation between married men and women for the number of years they spent at educational establishments – which may not be the same thing as intelligence. For example, among white, American, female college graduates aged between 35 and 44, it would normally be expected, assuming random mating, that only 10.7% of them would have married college graduates. In fact, according to a large survey, 59.9% had done so. At the other end of the scale, and again assuming random mating, it would be expected that of the same white female graduates 14.7% would have married men who had had an elementary education of less than eight years; in fact the proportion to have done so was 2.1%. It is easy to understand why this should be true: college students are, to a large extent, confined to each other's company and are likely to have similar interests, quite apart from any basic compulsion that may exist between intelligent people to marry each other. The effect of this form of strong assortative mating is to increase the genic variance of the next generation, to increase the range of intelligence and lessen the tendency towards uniformity.

At this stage it is necessary to refer to the phenomenon known as the regression to the mean. Where random mating exists, and where there is no evolutionary pressure favouring any characteristic that is controlled by many genes, the offspring will have a tendency to be nearer average for that characteristic than their parents. Should two fat people marry, their offspring will probably be nearer the average weight than they were at the same age. Even if parents are nearly of average weight themselves, the same tendency exists for their offspring to be even nearer the average weight. The reason why we have not ended up as uniform as tailors' dummies is that there is only a *tendency* to regress to the mean. It is only the most probable event. It is not a certainty. The various genes involved in a character such as weight will combine, from time to time and to confound the general rule, in a manner that is unexceptional. A child will then be heavier than both its parents. It will not have regressed. It will be the exception, as are all the very heavy, the very light, the very tall, the very short, and on and on.

When assortative mating exists, when tall people consistently choose tall people, there is still a regression to the mean, but less so. As a generalization, and if mating was random for height, one would expect the children to end up about half way between their average parental height and the average height of the population. (To get average parental height relevant to male offspring it is necessary to add six inches to the mother's height, and then divide both heights by two. For female offspring six inches should be subtracted from the father's height before the division by two.) In assortative mating between tall people there is still regression, but only about one-quarter of the way back to the population average. The irritating truth that such general dicta do not work out for any particular family is, of course, part of the inevitable contradiction concerning facts about an individual and facts about the population he lives in. There are rules, but each friend seems to fit none of them. There are generalizations, and everyone appears to be an exception. There are broad truths, and that is all. Tall people do tend to marry tall people more than average; their offspring, on average, are not as tall as the parents, but not as short as if that preference of tall for tall did not exist.

On then to sexual attraction, a most positive truth but a truth exceedingly difficult to pin down. The dance hall conversation might imply to any casual listener that only a limited proportion of those on view had any chance whatsoever of being married and mated. Both the belle of the ball, and the bull of the ball, might be considered the only possible procreators, with everyone else withering in frustration. As we all know, practically every girl on the dance floor does get married and is mated and is made pregnant. In the United States today, for instance, 85 % of all girls born alive will eventually produce offspring.

In other words, virtually everybody, big or ugly, wallflower or not, belle or bull, has a chance of contributing to the next generation, providing they survive to adulthood and do not have some genital defect. It would also appear – because there are no figures indicating otherwise – that the sexually attractive do not produce a greater share of that next generation. Admittedly one man's idea of attraction is not identical with another's, but there are some broad points of agreement, and yet no information to suggest that sexual attraction is important for the numbers of progeny.

Today is perhaps untypical; therefore what about the cave? Bearing in mind the fact that intercourse does not have to be outstandingly frequent in order to keep a girl fairly permanently pregnant, it is reasonable to assume that every cave girl was either about to be pregnant, pregnant or recently delivered of a child. So what about their consorts? Were they selected, with those more sexually acceptable serving a greater proportion of the available womanhood? One wonders. Perhaps sexual selection worked through the children. In times of scarcity a man might only bring back food for those women most pleasing to him. Their children would then be more likely to survive. Similarly the woman, in a time of scarcity, might give most food to the children of the man or men who pleased her most (provided of course that she knew which child had come from which man).

At all events, as Darwin pointed out and assuming that sexual selection does have evolutionary power, the qualities a man desires in a woman will eventually become more and more both of a female trait and a masculine trait. The exceptions are secondary sexual characters, such as female breasts or curves (controlled by hormones that a man cannot possess), but should he like long legs or hairlessness, and should more offspring therefore be delivered of longer limbed and less hairy women, the resulting boys and girls will be born with an increased proportion of these virtues. It was for this reason, said Darwin, that mankind in general has become less hairy over the centuries. Men still seem to prefer women without hair to their bodies and males will, in consequence, become less and less hairy themselves. Or perhaps in these artificial days, so different from life before or within the cave, the old powers of sexual selection in directing evolution have vanished. It would be nice to know.

By no means is sexual selection the only unknown in the business of mating. For example, and I am grateful to Harry Miller of Madras for the information, the children of the Muria tribes of central India are dispatched to communal dormitories at the age of nine or so. Sexual freedom is considerable, but the girls only become pregnant if they sleep with the same boy on more than one night. Partners are therefore exchanged nightly. There is a parallel here with the 'strange males' phenomenon discovered in mice by the British biologist Hilda Bruce. The arrival of a new male mouse will always cancel any successful conception by his predecessor and his conception will take its place.

As a second kind of unknown there is an observation made on kibbutz children. Isolated from their parents, and living in groups of their own age, the children are active in sexual play from early childhood, but they acquire inhibitions at about the age of ten. This tension disappears with the onset of adolescence and, although or perhaps because the children maintain strong emotional bonds with their fellow group members, they do not marry them. A follow-up survey of 2,769 kibbutz adults revealed that not one of them had married an individual from his or her particular group.

Marriage and who mates whom is critical to the next generation and to much of our current happiness. As we are now embarking upon so many new courses, with one-parent families becoming commoner (if present trends continue, one in four of all British marriages will have broken up irretrievably before their children have left school), with expectant fathers demanding paternity leave (the New York Board of Education has been petitioned that teaching fathers, now sharing so many maternal duties, should get the same four years' unpaid leave as teaching mothers), with artificial insemination influencing the procreational cause of marriage (see Chapter 16), with moral censure soon to be directed more at reproductive excess than sexual promiscuity, and with the binding importance of the family on the wane (in some parts of India, such as Andra Pradesh, a fifth of all marriages are still between cousins), with children both limited in numbers and conceived only when convenient, with the possibility of pills to affect the sex ratio, with the likelihood of selective abortion; no one can tell how these and other changes will have their effects, but effects they will undoubtedly have.

Mating in the recent past was geographically more limited, socially more restricted and medically more fraught than today. The impending changes are enormous and the likely results splendidly unpredictable. 'Why should I do anything for posterity?' says the adage. 'What has posterity done for me?' We are happily handing the future another generation with no more concern, or care, or anxiety than that with which we ourselves were created just one generation ago. It is a random business, oddly casual. It is genetical gambling. And long, one suspects, will it remain so. The doubt is whether it should a moment longer.

CHAPTER 4

Some experiments in human breeding

Pierino da Vinci – The *Bounty* men on Pitcairn – Friedrich's guardsmen – Hapsburg jaws and the Burgundian lip – George III and Gerrit Jansz

To any scientist the inevitable frustrations in the sphere of human genetics are intense. No breeding trials can be carried out, no tests conducted like those upon the mouse and *Drosophila*. In any case no quick human returns could be expected. Mankind's span of two decades between generations must be set against the mouse time of ten weeks and the fruit-fly gap of only a few days. The frustrations will continue – until such hour as society decides, in a manner unthinkable at the present time, to remove them.

However, there have been certain odd events in the past that, in one way or another, have been breeding experiments. From Florence in the Renaissance to Pitcairn Island, from the Prussian royal guard to royal families in general, and even to Afrikaners with sun-scorched hands, there have been some strange situations, all allied to practical human genetics and instructive in their differing fashion. For example, as a starting point, there was that odd and extraordinarily successful attempt to reproduce another individual in the mould of Leonardo da Vinci. It is perhaps stranger that countless other similar experiments have not been tried, and maybe there has been an insufficiency of genius to spark off identical thinking. At all events, Florence during the Rinascimento was a time and place for experimentation of every kind.

The experimenter, who put soul and body into the task, was Bartolommeo, Leonardo's half-brother. This man was twenty-two when Leonardo died (at sixty-seven) and, understandably, had been brought up in the shadow of the painter, sculptor, author, engineer and genius who was his elder brother. So great was Bartolommeo's veneration for his famous sibling that he contrived a scheme aimed, no more and no less, at his hero's re-creation.

70

Leonardo's father had been a notary, one of the Florentine gentry, while his mother had been Caterina, a peasant girl of too lowly a station to marry her lover. (Leonardo's birth is duly recorded together with the names of ten witnesses, but without the name of his mother.) Bartolommeo's plan, after he had sought out all available information concerning his brother's mother, was to seek out a girl whose attributes were similar. She had therefore to be young, to be a peasant and to be from Vinci, the village still standing even today in similar fashion not far from the modern Pisa–Florence autostrada. Bartolommeo found his suitable mate, mated her, even married her, and in due time was presented with a son, Pierino.

It is interesting, if the thing is judged as a genetic experiment, that no efforts were apparently contrived to make use of Caterina's genes via her family. After all, half of Leonardo's inheritance was from her. Bartolommeo, although a member of the family, had only 25% of his genes in common with Leonardo whereas his wife, the newly discovered peasant girl of Vinci, had none, so far as we know. Genetically it would have been better had he sought out a girl closely related to Caterina and perhaps, within the small village of Vinci, everybody was in effect related to everyone else. Bartolommeo's wife may well have had many genes in common with her antecedent, Caterina.

Anyway, Pierino da Vinci was brought up in the same Tuscan countryside that had nurtured Leonardo. According to Giorgio Vasari, a student of Michelangelo who achieved greatest renown as an art historian, the young Pierino 'displayed much grace, was beautiful with curly hair and of a quick intelligence . . . Unaided he practised design and made clay models, showing that the genius predicted for him was beginning to work. Bartolommeo therefore concluded that God had heard his prayer, feeling that his brother had been restored to him in his son'. The boy was taken to Florence at the age of twelve, put to work under Bandinelli, and then more successfully under Tribolo. By seventeen, and again according to Vasari, Pierino 'made everyone marvel and in five years had learned more than others do in a lifetime after long experience'.

Unfortunately Pier Francesco di Bartolommeo del Ser Piero da Vinci, to give the youth his full name, died at the age of twenty-three, having been born probably in 1530 (some say 1520) and dying in Pisa of a fever, probably in 1553 (some say 1554). So what of the experiment? Might he have become someone to be compared with his tremendous uncle? Vasari wrote: 'His abilities were so much admired that he was more in request than any man of his years, when Heaven decreed that he should go no further, and deprived the world of the excellent works which he would have produced had he lived.' In more recent times the art historian Ulrich Middeldorf has recorded that Pierino's works 'were mostly attributed to Michelangelo, and it is only recently that the real creator is slowly and in part receiving his due'. Antonina Vallentin, the Polish writer who has described Bartolommeo's eugenic conception in some detail (most writers skate happily over Pierino's origin), has written that: 'Chance or some obscure working of heredity reinforced the influence of the Leonardo legend and Pierino da Vinci became a sculptor of no mean quality. His short life, and the product of his epigonous gift, were like a last momentary flicker of glory of the dried-up Vinci stock.'

The experiment, therefore, was outstandingly successful. To be confused with Michelangelo is the kind of accolade that most artists would treasure beyond rubies, and to have reached such heights before attaining the age of twenty-three is also no small endeavour. After all, Leonardo lived to sixty-seven and when he was in his early twenties his principal renown had been extraordinary beauty and a nasty accusation of sodomy. Bartolommeo's idea was remarkable not only for its outcome but for the fact that he did it. Mankind has been fascinated by inheritance, by the presence or otherwise of similar ability in the succeeding generation, but has taken few steps to achieve a particular product. Bartolommeo actually tried out his own version of genetic reincarnation and, but for that Pisan fever, might even have succeeded.

Now to Pitcairn, the Pacific rock where another unique breeding experiment occurred with similar deliberation, however different the intent. Two groups of people, from opposite ends of the earth, sailed to this small, isolated and empty island, determined to welcome no intruders and start life anew. Not only was the event without precedent, but it was fairly well recorded. Therefore, as a further oddity of human genetics, it might have something to say.

When the men of HMS *Bounty* conducted their mutiny in mid-Pacific a prime cause, so it is alleged, was their long stay at Tahiti and their happy association with its inhabitants. Certainly the ship returned to that island the moment Captain Bligh and his supporters had been cast adrift in one small boat. Some of the mutineers then elected to stay at Tahiti but greater prudence caused the others to take their women with them and go elsewhere, British naval retribution being somewhat inevitable. When the *Bounty* sailed again from Tahiti, on 23 September, 1789, she carried 9 Britons, 2 Tubaian men, 3 Tahitian men, 1 Raiatean man, one babe in arms (a female who may well have been half-English, the *Bounty* having arrived at the island eleven months earlier), 19 adult women and assorted livestock. Alas, but the women and animals are less precisely detailed than the men. Alas, also, one woman leapt overboard almost immediately and then six 'rather ancient' women were left at Moorea, nine miles later.

Simple arithmetic clearly demonstrates the principal cause of trouble later on – fifteen men and twelve women – but from the genetic viewpoint the shortage is not important because each Briton always had a woman to himself. The six Polynesian males had to make do with a total of three women, but each man from the northern hemisphere had a mating partner from the other side of the globe. Such cross-fertilization had of course happened in the past, whenever opportunity had permitted, but on this occasion the hybridization was to be conducted in the isolated privacy of an uninhabited island, free from outside influence of all kinds and the Royal Navy in particular. At first sight, therefore, when on 15 January, 1790, the *Bounty* and her miscegenist community reached the two square miles of Pitcairn, everything augured well for a fascinating trial of insular inbreeding. Would there be harmful effects? Would basic incongruities from north and south be made manifest?

So far as the outside world was concerned the *Bounty* men, and their women, disappeared from view, seemingly for good. However, on their speck of an island, the experiment was in full swing. Discounting all other events which also took place on Pitcairn, such as murder and wife-stealing, the record of the nine British sailors in producing progeny was as follows:

Fletcher Christian, 26, Acting Lieutenant. Two boys and one girl by (probably) one wife.

Edward Young, 22, Midshipman. First wife nil. Second wife three boys and one girl. Third wife (Christian's widow) one boy and two girls. (Young was very dark and may possibly have been a half-caste West Indian – to complicate the racial story.)

William Brown, 27, Assistant botanist. A wife, but no offspring.

John Williams, 25, Able Seaman. First wife died within a year. He then took another from the three shared by the Polynesian men, to cause more dissent but again no offspring.

Mathew Quintal, 21, Able Seaman. First wife two boys, two girls and one infant death (at seven days). Second wife (previously Young's first wife) one boy.

John Mills, 40, Gunner's Mate. One boy (who died young) and one girl by one wife.

William McCoy, 25, Able Seaman. One boy and one girl by one wife (but she was the woman who brought the baby girl with her, perhaps McCoy's).

Isaac Martin, 30, Able Seaman. One wife (Jenny) but no children. (It was from Jenny that much *Bounty* information originates. Possibly unhappy through childlessness, she was the prime mover in the subsequent return to Tahiti.)

John Adams (who called himself Alexander Smith, following an earlier misdemeanour), 25, Able Seaman. First wife no children. Second wife three girls. Third wife one boy. (Adams/Smith was the last to die – in 1829.)

To sum up. Only six of the nine Britons fathered progeny, but they produced a total of twenty-one children to survive infancy, or twenty-two if McCoy's wife's extra child is added. There were five associations that were childless out of the fifteen *Bounty*/Polynesian unions to occur on the island, or 33%, which is a high figure if joint infertility is reckoned to be the sole cause. There were also two infant deaths out of the twenty-three children to be born from these unions, or 8.7% which is not a high figure for those days, let alone those conditions. The six Polynesian men are thought to have produced three, possibly four, children from the three women – reduced to two – allowed to them.

It was in 1808 that the mutineers' insular hiding place was discovered, after a silence totalling eighteen years. Of the original fifteen men only Adams/Smith was still alive, and of the twelve women 'several' were missing, but there were twenty-four (or twenty-five) children ranging in age from infancy to eighteen. The population had only increased from the original twenty-eight to about thirty-three, if those 'several' missing women are assumed to total seven, and of the nine *Bounty* names only five were to survive as patronymics – Adams, Christian, McCoy, Quintal and Young.

In short, as a breeding experiment, it could have been better carried out and should also have been immersed in secrecy for a period longer than eighteen years, barely a generation. With the hideout's discovery the way was open to both emigration and immigration. The childless

Jenny was the first to leave (in 1817) and nine years later Jane Quintal, daughter of Mathew, left – in pursuit of an English ship's officer. The first arrivals (in 1823) were John Buffett, to be the schoolmaster, an idea favoured by the islanders, and John Evans, who just jumped ship. Both these Englishmen married, with Buffett taking Dorothy Young, daughter of Edward, and Evans having Rachel Adams, daughter of John. By 1831, a mere forty-one years after the original arrival on Pitcairn, not only had there been a total of two emigrants and four immigrants, but the entire community, eighty-six in number, left the island for Tahiti, becoming eighty-seven en route.

Large numbers then died, others returned almost immediately to Pitcairn, but the seclusion had been shattered for ever. The experiment, if it can be viewed with such objective detachment, was over. It had all along been fraught with domestic strife – that lack of women had been disastrous. It had been too short. It had been confused by arrivals and departures, and it had then been pulverized by compulsory transportation and a subsequent splitting up of the varying descendants. The *Bounty* names still survive even to this day, but the *Bounty* isolation had been inadequate. Had the mutiny occurred a century earlier, and before the Pacific had become quite so crowded with European adventurers, a few generations might have survived in similar seclusion and thereby taught us a genetic thing or two about ourselves.

Perhaps the most authoritative genetic statements made in fairly modern times about the Pitcairn islanders have come from the anthropologist, H. L. Shapiro. His first visit was in 1934. On stature he found them slightly taller on average than the average for either of their two ancestral stocks. However, he found them shorter – by 5 cm. – than they had been in 1825 when the first generation were measured. Hybrid vigour can therefore, most tentatively, be proposed as the cause of their initial gain in height, a cause that diminished in its effect with the arrival of further generations.

On the results of their undoubted inbreeding Shapiro found it had 'resulted in no visible ill-effects, with the possible exception of the degeneration of their dentitions'. The bad teeth might have been the result of inbreeding, or the lack of dentistry, or the Pitcairn diet, or due to the mutineers' bad teeth in general and Young's in particular. Midshipman Young, who sired seven offspring (more than anyone else), had been described by Bligh as having lost 'several of his front teeth' at the age of twenty-two with the remainder 'all rotten'. He could be to blame for the modern decay. The most inbred individual discovered by Shapiro was 'a healthy islander showing no obvious stigmata of his restricted ancestry'. It had certainly been restricted; Christian appeared in his family tree 7 times, Young 6 times, Mills 3 times, McCoy 3 times and Quintal 3 times. In a normal tree, even one spanning as many generations, no ancestor appears more than once.

Other observers saw the situation somewhat differently, and perhaps as they had intended to see it. In 1898 the High Commissioner for the Western Pacific found the islanders greatly deteriorated 'owing to inter-marriage' and he indicated that the outlook was 'hopeless imbecility'. His successor, visiting the island in 1921, can readily be accused of greater objectivity:

'It might be expected that signs of degeneracy which were observed 23 years ago would be even more noticeable now; and certainly I saw none. Possibly former observers have been prejudiced by the theory that intermarriage over a period of years inevitably results in degenera-tion. This, I believe, is by no means the case when the original stock is sound.'

'It seems fair to conclude,' writes David Silverman, an Ohio lawyer who has made a hobby of searching out Pitcairn truth, that the island's community 'compares favourably in intelligence and adaptation to the human condition with much more advantaged communities in the main-stream of modern life.' Had they been more isolated and even less advan-taged for a much longer period the results of their inbreeding might have been made more manifest. As it was they did not suffer this affliction and, from the genetic point of view, the experiment in mutiny and privation apparently caused them no ill. This was good for them but, in the creation of knowledge, less satisfactory for the rest of us. Such events as the *Bounty* men's history have been rare in the past and are unlikely to happen ever again.

It might be expected that men in authority, desperate to have offspring capable of preserving that same authority within the family line, would have attempted to select women for their potentialities as if these mates were, so to speak, brood mares of impeccable stock. A basic understanding about the breeding of animals must have been widespread in the past and probably more so than today. Some of its methods might therefore have been transferred to the problem of human reproduction. Everyone knows, and knew, that strong horses are more likely to arise if strong stallions are mated with a stock equally rich, so far as strength is concerned, in its ancestry. Why then do we hear so little of this kind of selective husbandry in the human stables?

Not every royal person, for example, has married for love, or even for beauty, or even desire – at least of the conventional kind. Every other reason appears to have been the motive, such as politics (marry the daughter of a rival, or of a potential ally, or of a potential victim), money (any heiress), land (the girl next door), peace, war (Paris of Troy must have known the consequences when he took Helen) or simple convenience. Has any ruler ever taken a wife mainly for her breeding potential? Henry VIII, obsessed with desire for a male heir, for any kind of male heir, married his six wives for reasons, much simplified here, of love and politics (Catherine of Aragon), lust and a family's connivance (Anne Boleyn), love and relaxation (Jane Seymour), politics and foreign Protestant eligibility (Anne of Cleves), infatuation and further prompting of a powerful family (Catherine Howard), and finally companionship, sympathy and nursing (Catherine Parr). The kind of offspring was always irrelevant, save that it should be male, and the type of cunning needed by Henry's descendants was certainly a complex requirement. Perhaps Anne Boleyn had been the most cunning of them all, and certainly her daughter Elizabeth was among the wiliest of England's rulers.

Requirements were probably simpler in earlier centuries, when less cunning was necessary and more strength. Read of some warrior king of old and, as like as not, he stood taller than all the rest of his age *and* had valiant deeds to his name of severing men, or oxen, or anything else, at a stroke. In those days, when such elementary perquisites were also prerequisites, it would surely have paid to mate with some sledge-hammer of a woman and thereby produce appropriately solid offspring? But no; the men always strove for someone 'fair' or 'gainly' or 'of gentle bearing' and the poor products of their tryst had to make do with their refined inheritance.

However, there was one famous king fascinated with the genetic possibilities of gigantic stature. Friedrich Wilhelm, father of Frederick the Great of Prussia, measured 5 ft 7 in. himself but wanted a collection of the biggest men that could be discovered for his Potsdam Grenadiers. He had spies quartering Europe, bribing wherever possible and kidnapping when necessary. Money was spent with the enthusiasm of any great collector, provided only that the prize was sufficiently tempting. James Kirkland (or Kirkman), an Irishman in London, was given £1,000 sterling – a fortune – to travel to Prussia, and a further £200 expenses were paid to the discoverers and transporters of this single addition to the Potsdam force. Eventually Friedrich acquired over 2,000 giants, a substantial body of men who could hold hands over his carriage when marching beside it.

He even tried to breed from them. Large women were brought (and bought) into the fold to be married with the men, whenever possible. At this point the king came up against further harsh reality. Firstly, it all took such a time. Secondly, the matings suffered from regression to the mean, that genetic fact of life already mentioned which dictates that children of, for example, tall parents are quite likely to be half way in stature between the average height of their parents and the average for the population as a whole. Thirdly, many of his beloved soldiers were tall for reasons disconnected with their genetic inheritance, such as an error on the part of their pituitary glands, the endocrine organs that produce the growth hormone. Such individuals were likely to be poor breeders anyway, and certainly would not reproduce further giants, quite apart from that normal regression to the mean.

Friedrich Wilhelm was disillusioned. In any case, his own adult span – when he had power – was short relative to human generations, even had he possessed either more skill or better fortune with his regimental breeding. When he died, and when Frederick the Great took over, the Grenadiers were reorganized. Some of the brighter giants, such as the Irish Kirkland, became palace servants, but many of the tallest – and most feeble-minded – were merely sent away. As Nancy Mitford put it: 'The roads of Europe were covered with huge weak-kneed loons trying to find their way back home.' One more experiment had drawn to a close.

Now to the courts of Europe and their own royal variety of defectives. A suitable starting point is the year 1649 when Philip IV of Spain married Mariana of Austria. Presumably there was some rejoicing as bride and groom sped off to a palatial retreat and there, presumably, they initiated themselves into the business of rearing further members of their historic family. Instead of that rejoicing there should perhaps have been more solemnity because the marriage had a doomed nature to it, not so much in its personalities as in its genes. Any genetic counsellor of today, speedily spirited to that wedding, would have shirked in his duty had he not set fire to the altar in his efforts to stop the union.

The trouble was a strong streak of mental weakness in various of the couple's ancestors combined with a policy of inbreeding that had tended to concentrate the error. The spirited counsellor, had he been thorough, would have taken note of the parents of both Philip and Mariana, and of their grandparents and their great-grandparents and so on until he could probe no further. He would have been amazed at the replication of certain names among their joint antecedents. Had there been wedding guests (in fact the marriage was a quiet occasion) many of them would have been related over and over again to both bride and groom, and would have been hard put to it to say which side had their greater loyalty.

In a normal and unrelated marriage each partner can expect to have different ancestors. Each has, of course, 2 parents, and then 4 equally different grandparents, 8 different great-grandparents, 16 great-great-grandparents, and then 32 great-great-great-grandparents. A normal couple, marrying reasonably at random from a large community, could expect that his and her direct and grandparental ancestors would all be different people. Discounting the first two of any groom's generations (his parents and his grandparents) he should have $8+16+32$ direct ancestors in the third, fourth and fifth generations before him, a total of fifty-six people. So should his bride, making a grand total of 112 ancestors in those three earlier generations. For Philip and Mariana these 112 ancestral relationships were divided among only thirty-eight individuals. If put another way round this considerable duplication means that of Mariana's ancestors almost all were ancestors also to the man she had married. On discovering even this piece of news the genetic counsellor should have been shouting for a halt to the proceedings and reaching for his matches.

Worse still, and distinct from inbreeding, there had been mental illness in the family. For example, there had been 'the mad Isabella', and she had been grandmother to Juana la Loca, a probable schizophrenic who had grown increasingly hysterical, suicidal and melancholy throughout her long life. The importance of these two sick people from Philip and Mariana's point of view was the number of times their genes could have been handed down to them. Every single one of the couple's eight grandparents had Juana la Loca as a direct ancestor, and the name of mad Isabella occurs with even greater frequency in the family trees of all those involved. Every ordinary family must have mental illness somewhere along its line, if its current members are prepared to dig sufficiently far into the past, but the inbred families of Philip and Mariana had compounded this inheritance fearfully. From virtually every direction they could have received unsatisfactory genes, and the outlook for their children was therefore extremely unhappy. The counsellor would surely have advised them to stop their marriage, to look elsewhere for mates and dilute their inheritance with a bit of outbreeding, thereby reducing the odds loaded against their future offspring.

Such, then, is the background to Carlos the Bewitched, son of Philip and Mariana, born 6 November, 1661, twelve years after their fated marriage. (Six out of their previous eight children either had been born dead or had died very early. The infant Felipe Prospero had died, aged four, only days before Carlos had been born.) By the time of his subsequent marriage, to a niece of Louis XIV, the young Carlos was in bad shape. His jaw was so deformed that he could not chew, and his digestive system reacted to the arrival of great lumps of food by recurrent indigestion. The poor man also suffered from frequent fevers, from attacks of giddiness, from common discharges and from rashes. Mentally, as John Nada phrases it in his book about him, Carlos 'had a mind appropriate to the body'.

Eventually, in 1700, the deformed and deranged victim of his inheritance arrived at his death bed and became stone deaf. They put cantharides (Spanish Fly) on his feet and a freshly killed pigeon on his head to prevent vertigo. They kept him warm by placing fresh animal entrails on his stomach. He became speechless, almost understandably, as well as deaf and died fairly speedily. The post mortem discovered a heart allegedly the size of a small nut, three large stones in the liver, kidneys rich with water instead of blood, and putrid intestines. So ended, says Nada, 'the Hapsburg dream of conquering the world by marriages, and so died the last Spanish Hapsburg descended from poor old Juana la Loca'.

Carlos's terrible jaw was a feature of his Hapsburg ancestry. It had been recognizable for many earlier generations, long before Carlos had inherited a particularly severe form of it. For example, in a famous picture by Strigel of Maximilian I, his wife Marie of Burgundy and their family, the jaw is much to the fore. This was six Hapsburg generations before Carlos. Maximilian himself had a jutting lower jaw that could only have meant his lower teeth were in front of his upper teeth when his mouth was closed. Even closing his mouth must have been far harder to achieve than with the rest of us. In fact his grandson, the Emperor Charles V, more grossly affected with the same complaint, was told by a benevolent and straightforward Spanish peasant: 'Your Majesty, shut your mouth, the flies of this country are very insolent.' Maximilian also had thick lips, a long nose, and a long face, an unflattering assortment which Strigel has not disguised. Marie too had inherited similar disfigurements and the artist is just as faithful to the truth, or so it is easy to believe, for what portrait painter exaggerates facial misfortune? Her lips, especially the lower one, are also extremely thick. It appears as if she too is needing effort to keep her mouth securely closed.

The Hapsburg jaw and the Burgundian lip occurred repeatedly among the descendants of Maximilian and Marie. Dominant genes were probably at work, for dominance can usually be assumed when such a family trait appears so regularly, but the family's predilection for inbreeding further enabled this genetic affliction to be positive and visible. Not all dominant genes show their presence every time, but such genetic inheritance is more likely to show itself in an inbred line and the Hapsburg tendency towards their own kind was one of their greatest traits.

Take Philip II, for instance, who lived half way in time between Maximilian I and poor Carlos. In fact, take his four marriages, three of them consanguineous. One wife was a cousin, one an aunt and one a niece. Of his six children from that niece, five died in childhood. The sixth became the 'useless' Philip III, who married a cousin and thereby produced Carlos's father. We know how inbred he was (Philip IV), how inbred his wife was (Mariana of Austria), and how doomed poor Carlos was right from the start of his 'bewitched' existence. The jaw and the lips may not have done them much harm save for poor looks and a worse digestion, but they showed up the family's enthusiasm for itself. And that affection, very often, must have been the death of them.

This Hapsburg story, in demonstrating the working of a dominant gene, or perhaps one gene for the lip and another for the jaw, also shows the thinking and action of a royal line. It emphasizes the determination to keep out outsiders, to restrict the potential lines of danger. It is not always enemies that matter in the squabble for thrones, for kingdoms and huge estates; it can be relations. 'Beware of creating rival factions' might have been the Hapsburg motto. 'Keep ourselves to ourselves.' 'Inbreed never outbreed.' The policy may have helped to maintain them in power for several centuries; but, as Carlos the Bewitched and Juana the Mad both show, it did little or nothing for their general ability.

However, that lip and that jaw were not caused by inbreeding. Neither was haemophilia in the British royal family and its relations in Europe. The recessive haemophilia gene and the dominant genes for misshapen lips and jaws might equally well form part of other family trees. There are 2,000 males in Britain without the globulin factor in their plasma necessary for normal clotting. There are many of both sexes with jutting jaws and lips. And there is no evidence that families afflicted with such inheritance are any more inbred than is customary. With royal lines the trait is merely more famous. It is better recorded. It is not dispersed so rapidly, nor lost in the common crowd. The rest of us, largely ignorant of, say, our great-grandparents, may feel saddened at the realization that our great-grand-children will have no idea who we were and will not even know our names. So much for our immortality; but not so with royalty. They know their history.

Nevertheless there is one genetic story with a parallel between both royalty and an ordinary citizen. The king was George III of England and the commoner was Gerrit Jansz of Holland. These two men suffered from a similar complaint, or so it is believed, and in this story the commoner's medical state is better known than the king's. The affliction was porphyria, a metabolic disorder in which excessive amounts of the porphyrin part of the haemoglobin molecule are secreted in the urine. Not only is the urine, in consequence, a remarkable colour frequently likened to port wine, but the rest of the body can experience extreme sensitivity to light, much abdominal pain and a certain degree of neurological breakdown. George III is better known for his mental breakdown than for his urine, and the Dutchman is more famous for the blotched hands of his descendants than for his or their excretion; but both, or so it is said, had inherited porphyria.

Nothing is recorded of Jansz's family in Holland but he – together with his gene – stepped aboard a boat in 1685 and sailed for the new land of South Africa. There he bred and multiplied; his gene and its descendants did likewise. Three centuries after Jansz's immigration it was up to Geoffrey Dean, a doctor of Port Elizabeth, to do some detective work, to know the end of a story and to be curious about its beginning. The end involved the large number of current South Africans experiencing the symptom of 'van Rooyen hands', an inflamed and pigmented condition more prevalent in Dutch than British families living in that country. The Boers have one important similarity with royalty: a passionate interest in their forbears. Via the front leaves of their Bibles, they know who gave birth to whom throughout the generations. Dean was able, with the help of this written evidence, to discover the direct ancestors of all the Afrikaners and others now suffering from the Rooyen condition. A total of about 8,000 white South Africans as well as 2,000 coloureds (those of mixed Negro and European inheritance) are affected by it, but each family line that he examined went unerringly back to Gerrit Jansz. That one individual's single malfunctioning gene – only one gene is thought to be involved – had multiplied to 10,000 and will surely spread still further. It is not a lethal inheritance and Jansz's curiously immortal donation to his country is assured for all time.

Now to George III, the 'mad' king, the man who 'lost the American colonies' and was a resident of the British throne for precisely sixty years. (The only longer incumbent was, oddly, the genetic troublemaker of haemophilia – Queen Victoria.) George was not a perpetual raving lunatic, however much schoolboys may beg to differ, any more than Alfred and Canute idled their time burning cakes and failing to stem the tide. He was ill in 1765, when aged twenty-six, but this sickness was of little importance, however important the date to the American revolution. He was ill again in 1788, twenty-three years later, and this time he suffered abdominal pains, weakness, convulsions and stupor. Equally abruptly he recovered the following year, in time to frustrate Parliament from hastily establishing a regency in that revolutionary year of 1789. For twelve more years all was well, but in 1801 and 1804, when he was aged sixty-two and sixty-five, the same symptoms recurred. Once more the king recovered, but in 1810 the disease hit him again and this time there was no such alleviation. From the age of seventy-two until his death at eighty-one he fulfilled that schoolboy picture of the king, sick in body as well as mind.

The current suspicion is that he was a victim of porphyria. He did pass red urine from time to time, although the normal porphyriac passes urine that becomes reddish-brown only if left a while. How many of us would notice that fact in these non-chamber-pot days of the instant flush? Also he did have periods of insanity, although that is only an infrequent result of the disease. Perhaps kidney-stones were causing the bloody urine; perhaps something else was upsetting him mentally. Nevertheless, according to Ida Macalpine and Richard Hunter, recent investigators of the sick king's malady, 'the climaxes read like a text book case of porphyria'.

Earlier and ancestral illnesses help to clarify this picture. King James I of England suffered from colic 150 years earlier, but this did not prevent him casually describing his urine as the colour of a favoured Alicante wine. His mother, Mary Queen of Scots, is also believed to have suffered from the disease. From James, assuming the distant diagnosis is correct, the inheritance passed to both the Hanoverian and the Prussian royal lines. Later on the English imported a Hanoverian king, together with his inheritance, and in 1738 the genetic defect reached the infant who, twenty-two years later, ascended the throne and, as legend has it, promptly went mad and lost the American colonies.

It would seem as if two kinds of porphyria were involved in these two stories: variegate porphyria for the South African Dutch and acute inter-mittent porphyria for some of the kings and queens of Europe. So what did Mary Queen of Scots, beheaded by Elizabeth, have in common with Gerrit Jansz, Dutchman of Capetown? Much in a sense, for both are living yet. So are the *Bounty* mutineers. So too Friedrich's guardsmen. So too all our ancestors who produced us in their casual, feckless manner. But by no means was such inconsideration applied to their animals, the subject of the following chapter.

CHAPTER 5

The breeding of animals

Neolithic breeding – Dog origins – Domestic breeding –
Barking inheritance – Artificial selection – Robert Bakewell
of Leicestershire – Cattle – Fancied pigeons – Human
parallels

'*First Murderer: We are men, my liege.*
Macbeth: Ay, in the catalogue ye go for men;
　　　　As hounds and greyhounds, mongrels, spaniels, curs,
　　　　Shoughs, water-rugs, and semi-wolves are clept
　　　　All by the name of dogs'
　　　　　　　　　　　　　WILLIAM SHAKESPEARE

'*You can get beasts to weigh where you want them to weigh*'
　　　　　　　　　　　　　ROBERT BAKEWELL

It is possible, when within the arrogance of the twentieth century, to be amazed at the capabilities of much earlier generations of men. By and large we ourselves are entirely ignorant of the workings of modern things, of television tubes, transistors and nuclear reactors but, if the truth be admitted, previous and more fundamental discoveries made by simpler men are an equal mystery. How many of us, confronted by some eager tribesmen wishing to leap ahead of their stone-age toil, could make cement for them, or iron, or any metal or alloy, or glass, or even a brick that did not crumble before our eyes into the earth once more? Yet we reserve for our early ancestors, when we care to think of them, a sort of bemused condescension. We smile at their Neolithic simplicity, yet could not make even a flint arrowhead with one-hundredth of the grace that they employed. We smirk at those primitives who lived their lives perforce in caves and who, by some strange magic, not only painted exquisitely but used coloured dyes still active to this day.

Such conceit, on our part, makes it hard for us to realize quite how much had been achieved in the count-down of centuries before Christ came. It is arguable that more had been created in the way of crucial innovation before that date than after it. The frontispiece of *The Neolithic Revolution* (by Sonia Cole), for example, is a conjectural drawing of Jericho. It is harvest time. The men are cutting the einkorn, a kind of wheat. There are domesticated goats near by, and the field itself is irrigated by spring-water flowing along a ditch. The date of all this activity is 9,000 years before our own. There is not, but there could well have been, a domestic dog watching the scene. Moreover, it need not have been some vague mongrel of a pariah, but a creature of a breed and distinction fit for any hearth-rug of today.

Much like some well-aimed blow from a club, it should come as a jolt to our modern belief in the recentness of things that all the main groups of dogs were created long before Christ had appeared. In fact, they already existed when written history was making its first appearance, and when their various canine shapes and sizes were being made readily recognizable on stone and pottery. The breeding of dogs is extremely ancient, and modern breeders (who held their first dog show – in England at Newcastle – in 1859) have added little, whatever they may think, to the work done by the end of the Neolithic period.

There are several points of interest in the dog evolution. Firstly, although these animals are still happily able to identify one of their kind, there has been tremendous radiation in dog variety. There is curly hair, straight hair, wiry hair, silky hair, long faces, compressed faces, floppy ears, erect ears. There is a world of difference in the tail alone. As for weight, the chihuahua can be confronted by mastiffs forty times its size. Secondly, the bulk of this differentiation has been caused by man, deliberately and for different purposes. Man has controlled their random mating and has demonstrated the extraordinary variability inherent in the species (for it is just one species). Natural selection tends normally to enforce a greater uniformity among individuals but, given artificial selection or the artificial world of co-habiting with man, this variability can express itself. There is no reason why mankind itself, similarly freed from random mating and ready to reduce the persistent trimming of normal evolution, should not prove to be just as variable. Think of dog variety, and then think of the human frame in a similar range of livery, of shape, of size and of ability. It could happen.

The dog story is also, to some degree, a hastened version of mankind's own history. When men were banding together to form large village communities they brought their dogs with them. Since then both have experienced, in similar fashion, the cultural and environmental changes that have occurred – canned food, heated rooms, soft furnishings, urban compression, disease control and increasing unnaturalness. At the same time dogs, with their quicker breeding cycle, have passed through some 4,000 generations while man has experienced a tenth of that number. It is these facts, according to John Paul Scott and John L. Fuller in their book on dog behaviour, that suggest a hypothesis: 'The genetic consequences of civilized living have been intensified in the dog, and therefore the dog should give us some idea of the genetic future of mankind ... In short, the dog may be a genetic pilot experiment for the human race.'

There is one further important point. Evolution, or rapid genetic change, happens fastest when a population is divided into small isolated groups that have occasional genetic contact with each other. This was precisely the condition of mankind 10,000 years ago when dogs were becoming a part of each isolated scene. The trading, or warring, or friendly association between each group permitted a degree of genetic involvement with the dogs next door. American Indians, for instance, with living areas well spaced apart, each had their own breed of dog. These breeds were not greatly different from those of each neighbouring tribe, but across the continent the differences became intense. So too in earlier times with all the fairly isolated, semi-independent groups of people that were the ancestors of us all.

Then came the Neolithic revolution. During that time, that tremendous upsurge when mankind took off in its present direction, the various breeds of dog were dotted everywhere. They had been nurtured in isolation, and were then ready to accompany their migrating masters, to be sold or captured along the way. Practically every dog name is based upon the geographical area from which that creature had arrived. The spotted dog, for example, may not have originated in Dalmatia (in fact it was India) but it came through there and that was good enough for its new owners. So too with the spaniel (allegedly Espagnol), the greyhounds (allegedly Greek), the saluki (thought to be Seleucid).

The relative gentleness of mankind's movements became transformed when the age of discovery began. No longer did small groups wander along conventional lines but boatloads leap-frogged among the continents. Everyone took dogs along with them and they brought dogs back. There was much panmixis, or wide-scale interbreeding. Sometimes this was successful; sometimes not. New breeds were formed, often at the expense of the native varieties which died out, as in America, as in South Africa.

At this stage the human comparison becomes more intense. Native human breeds have been extinguished; think of Tasmania, think still of South America. And there has been interbreeding, creating new varieties and colours of men; think of the West Indies, of South Africa, of Hawaii. In the case of dogs the wish to preserve and intensify certain characteristics has been pursued vigorously. There have been some ghastly errors, such as over-long dogs whose abdomens trail the ground; but, in the main, the breeders have acquired what they consider to be desirable characters. Therefore, as the dog story has so many parallels to our own it is worth while spending a little more time on its history in order to probe, perhaps, into our own future. Examine the dogs, therefore, but think about men.

The Canidae were derived fairly recently, late Pliocene and early Pleistocene, from more primitive and ancestral predators. They fanned out rapidly into different regions but never lost their ability to adapt. (Neither has man who has fanned out just as far.) Among the Canidae were the wolves (*Canis lupus*), a species having much in common with early man that exhibited co-operative behaviour when attacking its prey. Wolf and man were direct competitors and this similarity of life style may have had much to do with their eventual union. It is believed that all modern dogs derive only from the various races of the wolf species and that the association of wolf and man leading to dog happened again and again: American Indians were accustomed to taking wolf cubs frequently, either to amend their dog stock or to start afresh. Probably this process of turning wolves into dogs was spurred on by the considerable growth of forest when the ice retreated, and when man's extreme inability to smell out his prey in that tangle of hiding places became a terrible disadvantage.

Anyway, the dog, a remarkable social invention, was the first domestic animal. (Some say the reindeer won, but it is difficult to exploit reindeer without the aid of dogs.) The new accomplice could be tamed, if caught early enough or reared in human company. It could hunt. It could herd animals. It could protect and give warning, although wild Canidae are not given to protective barking as much as the tame varieties. Every tribe must have wanted this new aid to living, and good bitches must have been tremendously in demand.

However, a good forest dog is not necessarily suitable either for the open plains, or for herding semi-domestic stocks, or for guarding the encampments. Therefore there had to be selective breeding for specific purposes. This is thought to have made much of its advance between the Palaeolithic and Mesolithic times, or between 20,000 and 10,000 years ago.

Recognizably different breeds of dog have been recovered from archaeological sites of that age, notably in more northerly regions, and by the time the Mesolithic was yielding to the Neolithic there were hunting dogs, sheep dogs and even the 'toy' dogs of the Maltese type. Living with the early Egyptians there were at least five major kinds: the basenji, the greyhound, the Maltese, the mastiff, and a sort of chow.

It is important to appreciate that the changes, however striking, were not basic. They were more of degree, of proportion. Take an ordinary dog, for example, one neither small nor large in any direction, and place it in front of a fairground's set of distorting mirrors. The dog may bark with instant disquiet, but it is possible, with suitable juxtaposition, to make it whippet-thin, or spitz-faced, or as massive as a mastiff. At the same time the observing human being can regard himself, thereby learning some of the possible variations equally feasible with his body – perhaps a long trunk and short legs, or a thin neck with a tiny head, or a sleek whippet of a frame. Of course dogs also differ in those characteristics that were different even in the basic wolf, such as colour, such as temperament, such as wiriness or smoothness or furriness of hair. With man too there is great variability, either unsuppressed or strengthened by natural selection, such as hair form and colour and personality.

A large number of the dog differences are the result of neoteny, the retention of juvenile features. This process occurs when any creature stops growing or achieves its adult state while still possessing characteristics previously associated with its more immature form. Puppies, for example, have silkier, less wiry, hair. Their ears, in general, are floppier, less straight and less erect than with the adult. Their tails are more likely to droop, less likely to stay firmly away from the legs. There is no law which states that these features must be linked with the juvenile form but some of the dog changes have inclined towards these puppy-like features. (Many of the Pekinese characters are an example.) Human beings are also believed to demonstrate some neoteny; their skin is thought to be smoother and more baby-like than before; their general hairiness reduced; their supraorbital ridges, those bony eyebrows of early man, less pronounced. There is no reason why further neoteny should not take place, or even be promoted artificially. We could then become still less furry, more like South American Indians who grow no facial hair, more like our children, with their different proportions, their smoother skin, their softer hair.

Neolithic man lived in different environments, and his dog requirements therefore differed from place to place. These have also changed in more recent times and the breeders have had to adapt the supply. The big mastiffs have been made yet bigger, and built for war. Pet or toy breeds, suitable for living within a household, have been more favoured. In the Aztec world dogs were bred for eating, for their hair (as wool) and for their role as beasts of burden. The whiteness of the English terriers was a requirement for greater conspicuousness within the undergrowth. Even the arrival of breech-loading guns demanded a different animal; the steadfast and entirely patient behaviour of the pointer, doggedly stationary while his master fulfilled the ritual of preparing a muzzle-loader, was far too steadfast when a more capable firing-piece became available. Therefore setters were deliberately bred from spaniel stock to sit or 'set' when the quarry was detected, and then to move forward when the quicker gun was ready.

In short, different dogs have been our steady requirement. The breeders, whether putting their animals out at night to be mated by the wolves, or actively preventing any such event, or trading particular animals up and down the migration routes, have always been attempting to satisfy most particular needs. During the past 20,000 years these men have been outstandingly successful in unleashing the potential variation in the domestic dog, the animal that once was wolf. There is no reason whatsoever to believe that the wolf/dog of old had any greater inherent variability than any other species such as, for example, mankind.

Not only have physical characteristics been selected for; behaviour has always been important. The breeder has had to provide for an animal's ferocity (guard dogs), or overall friendliness (general pets), or individual friendliness (loyalty to one person alone). There is, for example, the matter of barking. Some individual dogs bark more than others; some breeds do. It can even be proved, and Scott and Fuller have done so, that this behavioural trait has a genetic basis.

For their experiment they took the basenji, a poor barker, and the cocker spaniel, a noisier creature. Basenjis not only bark rarely, but make less noise when they do so and stop quicker should they start. Presumably the barking facility served less purpose in the African forests, or was pounced on more readily by dog-loving leopards. Whether their masters selected them for their quietness, or whether leopards saved them the trouble, the modern examples of this African breed howl, or yowl, more than they bark. Any kind of howl, with its wavering note, is harder to locate and pinpoint than a bark. The cocker, on the other hand, named after its ability for flushing out woodcock, is a standard barker. Given a form of incentive to do so the cockers barked, during the experiment, on 68% of these occasions. The basenjis not only barked less loudly and more briefly when subjected to similar stimuli but responded on only 20% of the occasions.

This behavioural characteristic, bark-a-lot versus bark-a-little, was therefore a most suitable trait for investigation. The experimenters' plan was to cross cocker with basenji, to cross again the offspring of this mixed mating, and then to back-cross the second generation both with pure cockers and with pure basenjis. At each stage the willingness to bark would be noted.

The eventual result of all this breeding, and of all its measured noise, was clear. Whereas pure cockers barked, as already mentioned, at 68% of the stimuli, the basenji/cocker crosses barked at only 60% of similar provocation. By the time those offspring had been crossed with each other the willingness to bark had been reduced to 55%. But then came the back-crosses. When the second generation (of mixed basenji/cocker stock) was mated with pure cockers the result was a 65% proneness to bark. The other back-cross (with pure basenji) resulted only in a 50% proneness. In other words, genetic inheritance is playing the crucial role in this particular piece of behaviour. The reluctance or otherwise for these dogs to bark at a certain stimulus lies in their genes.

Humans also have such behavioural traits, whether as individuals or as groups, and these too can be genetically based. (For a time it was not infrequently stated that heredity had no effect upon human behaviour, but this was during the period of widespread revulsion against the exaggerated claims of the early eugenicists.) Which language we speak is of course environmental, but the latent ability to speak is genetic. Therefore, although it may be simplest to think first in terms of breeding physical characters, of height, eye colour, limb shape and proportion, it should never be forgotten that personality, manner and other features of our behaviour are just as involved. They too can be selected for, or against. It is possibly more comforting to imagine that our own individuality has come to us in some mystic manner, as a spiritual creation outside the normal laws; but it has done nothing of the kind. It is as genetically based as the rest of our frame, and therefore just as susceptible to the laws of breeding.

By way of general amplification it is necessary to examine other breeding successes of the past. Every domestic creature and every crop has been subject to selection by man. Some idea of the deliberation involved is achieved by looking at the wild ancestors, the minute originators of the fat cereals of today, and at the miniature and bitter Bromeliads, for example, that have been swollen into the modern pineapple. The optimism with which these early varieties were chosen, husbanded, harvested, and then selected again and again for their particular merits is astonishing. These forerunners of our current foods were pathetic by comparison with the fruits, and seeds, and tubers, and roots that we enjoy today. Some, such as sweet corn, the fat, swollen and golden *Zea mays* of today, cannot be identified – or at least there is no agreement – with their diminutive forbears. Stone-age man should everywhere be given more credit than is normally his due because, as Sonia Cole puts it: 'Every food plant of major importance to mankind was grown in the Neolithic stage of culture, just as all the main animals reared for meat today were domesticated during that period.' It was an extraordinary age.

More extraordinary still, the Neolithic products were hardly improved upon until quite recently. One name then stands out. One man conducted a breeding revolution almost on his own, just at a time when everyone was ready for it. Robert Bakewell, of Dishley in Leicestershire, the rich agricultural county of central England, inherited his father's farm in 1755. He was fascinated in the practice of breeding, believing, as he repeatedly said, 'that you can get beasts to weigh where you want them to weigh'. The new enclosure laws of the eighteenth century were inhumane to many, virtually killing off the yeoman freeholders and turning villagers into landless labourers, but they were a boon to the efficiency of agriculture in general and to breeding in particular. The random mating of animals sharing communal grazing ground became controllable wherever the open field system of medieval agriculture was abolished. With the possibility of deciding which male should most profitably inseminate which female, the occasion was immediately ripe for breeding advancement. And Robert Bakewell was there with the right ideas at this right time.

In his view the quality to be selected for in cattle only concerned the valuable joints. Everything else, such as the head, neck, leg, horn and colour, he regarded as non-essential, however much these features had been prized by fellow farmers. He also exploited a new policy. The normal method of achieving better stock was to cross two alien breeds. Bakewell felt that, as one of the two breeds was bound to be inferior to the other in any desirable quality, such a bastard practice merely adulterated the better breed. It was plainly preferable, or so he thought and then carried out in secret, to mate animals that were of the same line of descent, and even of the same family. If one beast had a certain quality that same quality was more likely to be either conspicuous or latent in its close relatives. Therefore, instead of breeding outwards, he bred inwards, and with staggering results.

He produced, for example, a new and Leicestershire breed of sheep. Within fifty years it spread to every part of the United Kingdom, to Europe and to the burgeoning United States. In 1760 he had hired out his rams for a few shillings a season. By 1770 they were fetching twenty-five guineas a year, and in one twelve-month a particularly famous ram named Two-pounder earned him 1,200 guineas. It was claimed, as a result of that animal's work, that England 'had 2 lb. of mutton where there was only 1 lb. before'.

In cattle he was not quite so successful, in that his products were better beef than milk producers, and people wanted both; but he had an excellent breed of black horses, a team of draught cows, an irrigated farm giving him four crops of hay every year, and much else besides. Visitors flocked to him. He entertained Russian princes, royal dukes from France and Germany, British peers, and agriculturalists of every degree. He showed them his 440 acres. He paraded his extremely docile animals (another feature he selected) but he never told the sightseers a thing about his methods. In fact, when the extreme age of some of his best breeders forced him to sell them, he is said to have infected these animals 'with the rot' in order to prevent any further service from them.

The visitors must somehow have acquired greater wisdom because that eighteenth century witnessed more progress in breeding methods than had occurred since the Neolithic. At Smithfield, London's meat market, and in 1710, the average weight of a bullock (after walking there) was 370 lb. By 1795 the animals were still walking, but weighing an average of over 800 lb. Bakewell died that same year, allegedly in poverty, and with, as Lord Ernle phrased it (in *English Farming Past and Present*), 'his own weapons turned against himself'. Native animals of other districts, which had been improved on his principles and despite that secrecy, began to hold their own and then to overtake the famous Bakewell breeds.

All the ancient breeders, whether Neolithic, eighteenth-century or even nineteenth-century, were working by rule of thumb and not by the science of genetics. Bakewell and all the others, however cunning their various schemes, were entirely ignorant of the Mendelian laws, and no less so

93

than those practical Neolithic men of Mesopotamia and elsewhere. Even now that genetics has arrived the art of breeding is still somewhat hit and miss, with rules of thumb firmly operating, with hunches playing their part, and with genetics being introduced only from time to time. By no means has the agricultural problem – of getting more product for less money in a shorter time – been solved, and Bakewell could leap again into the task with as much fervour as before. For instance, he would feel entirely at home with the following statement, culled from a recent report: 'Bulls had daily gains 13.4% higher than steers and a feed cost per lb. of gain 10.6% cheaper. The bulls were slaughtered 19 days sooner than steers and produced carcases 17 lb. heavier . . .' Should, in other words, Britain and certain other countries persist with their (apparently un-economic) custom of castrating bull calves?

There is also current enthusiasm over the enormous 'draught' cattle of the European continent. The Charolais, never imported into Britain before 1961, have the highest 400-day weight of any breed ever recorded in the United Kingdom. When crossed with commercial suckler cows, they produce faster-growing calves and bigger adults than Britain's big-gest breed, the South Devon. They are also superior to the classic types, the Hereford, the Beef Shorthorn, the Aberdeen Angus, and where would Texas be without the Hereford? Heavier still, and perhaps even more entrancing to the reborn Bakewell, are the Chianina of Italy. The bulls can stand 6½ ft at the shoulder and weight 4,000 lb. (Remember those Smithfield bullocks of 370 lb., mere dwarfs to these giants.) It is expected that Chianinas, should they be imported into Britain and other suitable areas, will produce one-ton two-year-olds having lean carcases of over 1,000 lb.

Bakewell was a poly-breeder, interested in every species capable of turning mere food into faster profit. He would therefore be delighted to know of today's endeavours to make sheep more like pigs. As everyone knows a sow can produce large litters more than once a year; a normal ewe, by comparison, is pathetic in creating but one lamb, perhaps two, in the same twelve months. However, some sheep have produced five lambs per litter and could – in theory – be induced to do this twice a year. At present these prolific breeds have poor carcase qualities. So should normal ewes be injected with gonadotrophic hormones to increase their ovulation, however unpredictable the size of the litter? (Some previously infertile human mothers have recently been producing remarkable litters of offspring as the result of hormone injection. The technique, whether for women or ewes, is not without its difficulties.) Or should the offspring of all ewes naturally producing larger litters be selected, despite the long time until it is known if the selection procedure is bearing fruit? Once again, despite the arrival of genetics three-quarters of a century ago, despite even the word biology coming since his day, Bakewell could make himself thoroughly at home. The tasks of animal husbandry are as problematic, and as intriguing, as ever.

It is often argued that mankind has a basic longing, partly suppressed by urban living, for growing things. A general desire for acquiring earth beneath the fingernails is totally understandable, bearing our long agricultural dependence in mind. So too is the long-standing wish to chase things, to aim with weapons and to succeed. It would appear that breeding, also firmly entrenched in our history, has an equally basic grip upon us. Anyone with a dog purer than some urban pariah is happy to learn about its forbears and to be deliberate, insofar as oestrous compulsions make this feasible, to control its descendants. Tropical fish are subject to similar interference in their domestic lives, given the cyclopic interest of their masters beyond the tank. Half the fun of following horses, and of staking money, is knowing by and out of whom a particular mount has come to pass; the bets are placed upon an ancestry as much as current form. To keep rabbits is to breed rabbits, to know that a Polish weighs some 2–3 lb. while the Flemish giants can swell up to five times that size.

There is also pigeon-fancying, as strong a drug as any. It is possible that the pigeon was second (or even first) to the dog in providing tame companionship for man. Some pigeons are still cave-dwellers and so was man. All the modern domestic breeds – there are over 100 – can be traced back to the ancient civilizations of India, Persia and Asia Minor but no breed is indigenous to Europe or America, however strong the current cult in these two continents. Moreover, the homing instinct exists in all pigeons and would be of little appeal to nomad tribes wandering, as did so many of our ancestors, in tune with the seasons. It would be of greater interest to stationary people, and the first real settlements were in Asia.

For breeders the pigeon has excellent qualities. Wilson Stephens (writing in the *Field*) summed them up as 'hardy, clean and monogamous'. The last asset is particularly desirable. Not only do they have a faithfulness to their mates that cannot be emulated by other domestic creatures (think momentarily – for that is sufficient time – of dogs, or of rams, or of bulls, or even of humans) but this loyalty can readily be disrupted by the breeder. He can therefore eat his cake and have it. He can know that a certain partnership will survive and breed true, or he can replace one mate with another and be rewarded for his perfidy by further loyalty on the pigeons' part. As Stephens put it: 'They modify their amours with a readiness which Mr Barrett of Wimpole Street would have approved in his daughters.' Finally, there is a most satisfactory punctuality in pigeon affairs: ten days after the mating the first egg appears, two days later the second egg, three weeks later the hatching, and so on to maturity. A breeding pigeon programme can be planned as if the biological materials involved are part of a production line.

The result, apart from there being a considerable body of enthusiasts, known collectively as 'the fancy', is that more domestic breeds of pigeon exist than there are pigeon species in the wild, but all of man's birds come from just one species, *Columba livis*, the rock dove. Would the pigeon or the horse, or the dog, be quite so fancied if they were not so adaptable to mankind's predilection for breeding things? I think not.

It is not impossible to imagine that mankind may one day apply the enthusiasm for breeding other species to humanity itself. Biological advance will steadily make it easier for him to do so and greater genetic knowledge will provide better guide lines. By then, to revert once again to the dog, that species fashioned by man into the most blatantly diverse creature of them all, canine variation should be still more extreme.

As of now the tendency towards diversity has gone to greater lengths in the dog than in man, and for two reasons. Firstly, natural selection for the dog has been relaxed for a longer time – in terms of generations – than for man. Secondly, artificial selection – of the dog by man – has deliberately preserved some unwelcome canine mutations merely to increase that diversity. Has the dog species therefore suffered genetically? Natural selection is impartial, favouring no species in particular, but has its lack of influence caused some kind of lack and some kind of unfitness in the dog? By the same implicit token, does our current artificial world render even humans less adequate for its artificiality?

The answer is that the dog does not seem to be genetically weak. For example, the current breeds are largely more fertile than their wolf ancestors. Wolves mature at the age of two, and produce four or five cubs per annual litter. There may be selective reasons for this casual growth, such as the inability of the habitat to support large numbers of predators, but the wolf's relative infertility is still the case. The dog is, generally speaking, sexually mature before the age of one and can produce two litters a year. By this criterion man-handling has not harmed the dog.

Nor has it in the range of capabilities. As with humans, variety is the keynote and today's dogs have broadened all the old wolf characteristics. Terriers, to quote Scott and Fuller again, are more aggressive than their wolf forbears; the hound breeds less so. Greyhounds are faster than wolves; short-legged dogs, like the dachshund, less so. The good scent dogs are better trackers than wolves, while terriers are poorer. Sheep dogs can herd more effectively; other breeds would not know where to begin. (Think, very briefly, of the pekinese.) Many game dogs are more interested than wolves in game, reasonably enough, and share man's enthusiasm for birds.

However, as in the pentathlon, there is no dog to exceed the wolf in every degree. Each gained dog ability has been achieved at the expense of some other virtue; the steady, scent-orientated fixation of the blood-hound, for example, makes one fear, possibly for his intelligence, certainly for his nimbleness and general group co-operation. The various losses are important. It is doubtful if any breed of dog could survive in a wilderness where the wolf is in competition.

In fact dogs have only gone wild where there is no such rivalry, as on the Australian continent. The aborigines arrived in that empty land during the Mesolithic age, and are thought to have brought the dingoes with them, either tamed as part of the joint community or semi-parasitic on its fringes. On that continent there had been a marsupial wolf, *Thylacinus*, surviving poorly in the forests, but the subsequent arrival of the white man helped to put paid to its modest viability. The result of both events, of both white and black invading the brown landscape, is that yellow dog dingo is now wild and howling at the moon. As dogs go, this one's attributes are more general than specialized and it has made effective use of all of them in conquering the continent.

The lesson for mankind is that natural selection does not favour the creation of some super-beast, highly able in every degree. Either the result is a good all-rounder, like the wolf, able to track, to scent, to co-operate, to run, or it is a specialist, closely adapted to the needs of some precise habitat. Man took the all-rounder wolf and turned it into a whole variety of dog specialists, each excellent in something but always paying the price by losing in something else. The lesson is that this same ruling applies to no smaller degree with humans. All those mythical Greek heroes were strong *and* cunning *and* beautiful *and* fertile *and* wise and generally long-lived. The dog story, more down to earth, would have us believe otherwise. Beethoven and Leonardo and Alexander all have to be separate people. They were brilliant in their own manner but also, respectively, deaf, homosexual and short-lived; they could not excel in every particular. Only society can be super-human, as wide-ranging as the variability within it, as excellent as all the peaks of brilliance within its compass. To breed wisely, therefore, is to increase the range, to treasure the variation. It is not, as Heinrich Himmler wrongly supposed, to choose a particular ideal and aim alone for that.

CHAPTER 6

Human inheritance

Mutations – General inheritance – Baldness – Hair whorls –
Sex chromosomes – Sex and autosomal linkages – Defects of
the hand – Peculiarities of the eye – Blood groups –
Paternity – Deafness – Moles – Hair – Dwarfism – Ears –
Twinning – Genetic diseases – Congenital malformations –
Ante-natal diagnosis – Spina bifida – Mongolism – Epilepsy
– Diabetes – Sickle-cell anaemia

'Take care to get born well'

GEORGE BERNARD SHAW

*'Their faces were not all alike, nor yet unlike, but such as
those of sisters ought to be'*

OVID

At last, in a sense, to some facts about human heredity. Nevertheless it
would have been improper, or so I believe, to have encountered them
earlier, to have omitted the introductory material. A human being makes
more sense if he or she is seen as a product of evolution, of earlier times,
of a different world. So too with our inheritance. It did not arrive as a
bolt out of the blue but stands upon its ancestry. Hence it is appropriate
that this chapter should appear, apparently belatedly but on a firm foundation taken mainly from the past. However, even now, there is further
delay as a few more introductory statements are necessary.

The creation of any human being is the accumulated result of haphazard
errors from the past. No species can become distinct from any other species
save for mistakes in copying the genetic material. These have made us what
we are, and continue to make us, again and again. It is not as if there is
some blueprint, with *Homo sapiens* inscribed at the top, that was drawn
with us solely in mind. No one and nothing had us in mind. 'We are,' as
Julian Huxley has said, 'just as much a product of blind forces as is the
falling of a stone to earth or the ebb and flow of the tides.' We have

just happened, and flesh was made man by a long series of beneficial accidents.

The intricacy of DNA, the complexity of its code and the orderly efficiency of the various forms of RNA, make it easy to assume that any error anywhere along this bewildering line will surely be a major disruption. Like a child released within a cockpit it is hard imagining that anything but harm will result. So too with mutations, the errors of the system. They are harmful, almost always, but on occasion their results can be irrelevant or even good. The boy's actions on the flight deck will be mainly detrimental but some will be inconsequential and a few may even prove to be of benefit.

In imagining evolution it is easiest to think of end results, of the lion growing stronger and the antelope becoming yet more fleet of foot. Plainly either change is advantageous. It is infinitely harder to think of these amendments in terms of an atomic adjustment within DNA, of a minor alteration that leads, via an alteration in the manufacture of a protein or even a change in its timing, to a stronger body or longer legs. Surely, one argues, any such change cannot occur entirely independently? What happens to the liver, or the heart, or the pancreas should there be a move towards longer legs? What indeed! Do not forget for an instant that the business of growth involves the development of one cell into (for human beings) one hundred million million cells, but that even more cell division is linked with replacement rather than growth, with a complete turnover of blood, for example, every single month, with a perpetual manufacture of the skin that clothes us all. It is not so much inconceivable that mutations happen; it is more incomprehensible that their new effects can ever be absorbed into the system, and sometimes beneficially.

Nevertheless, for all the near-inevitability that they are disastrous, evolution could not have proceeded and cannot proceed without them. They are the means of change and natural selection is the promotion of their successive transformations. Not until 1927 did anyone learn how to cause mutations when the American geneticist H. J. Muller clinched this point after he had bombarded *Drosophila* flies with X-rays. It was a momentous piece of work (with the Nobel committee rushing him their prize eighteen years later) partly because it promoted genetics by permitting new mutant strains to be created and also because it made humanity aware of this potential threat to all inheritance. With the official start to the nuclear age in 1945, and with the knowledge that ionizing radiation could form mutations, the world was suddenly alarmed to realize that all future generations could be harmed, or even prevented, by current mis-demeanour. The list of chemicals, radiations and even temperature changes that are now known to have the capability of inducing mutations is of formidable length.

Before leaving mutations, it is also vital to realize that most inherited errors of metabolism in humans are not manifested in the heterozygous state. In other words, unless the genes representing these errors are received from both parents (to make the homozygous condition), the error's effects will probably be masked by the proper workings of the good gene. Such so-called recessives can only show up when there are two of them and, for genetic disease, most of the causative genes have been found to be recessive. If they were dominant, and manifested them-selves in the single or heterozygous condition, they would arise in 50% of any carrier's descendants and the affected gene, assuming it to be harm-ful, would die out more speedily under selective pressure. If it is recessive there is only selection pressure when the two genes combine in one indi-vidual. Already 1,500 genetically caused diseases, almost all of them recessive, have been listed for man and they – however unwelcome to the victims – have been the blessing of human genetics.

It is unfortunate that human genetics has to be so allied with mal-formation rather than with more fortunate inheritance. After all, even if coupled with some hideous congenital abnormality, the rest of the body is more than 99% normal; but it is the very nature of the abnormalities, whether dominant or recessive, whether they only arise in one sex or the other, whether they can skip a generation, and whether their effect is always complete or only partial, that provides all the genetic interest and elicits all the information. To be normal, to have two eyes and all your teeth, to have two ears and all your hair, is to be extraordinarily uninter-esting; but to possess hairy ear rims or a thirteenth pair of ribs or webbed toes is to have the geneticists flocking round as if you have just invented that better mousetrap. In fact, the word 'monster', a general term these days for all kinds of severe congenital abnormalities, comes from the Latin 'monstrum', a sign. Such freakish events were believed in the past to be a divine portent. Today they are the ill wind that blows good infor-mation the way of all geneticists.

Perhaps it is as well to be warned that exceptional interest in either antecedents or descendants is not always either wise or rewarding. It may be interesting to trace a dominant feature back and forth but the whole truth can be uncomfortable. In one British survey, dedicated to another purpose, remarkable facts about infidelity were unearthed. The research involved the rhesus factor, that blood-group awkwardness which poses a problem for second-born children (and their parents), and the research necessitated the taking of blood from all three principals, the mother, the child and the father. It was routine work, with the various blood groupings being variously tested, but it was quickly realized and then precisely reckoned that 30% of the children could not have been sired by their alleged and loving fathers. The actual proportion of wrongful attribution must have been even higher because some of the children may have possessed blood groups that corresponded *by chance* with those that would have been donated by their legal fathers. The area of this survey, for those of you reading these words with mounting dismay, is pinned down no further in the records than 'a suburban district of London'. Science can keep faith with its patients, even if they themselves are less than loyal with either their nearest or their dearest.

As one further point before entering the inheritance lists, that Mendelian simplicity of dominant and recessive genes can be much obscured by the events that actually happen. There are such complexities as partial dominance, poor penetrance (by a gene) and other intermediate attributes. Here is a quotation from one textbook, attempting to clarify this particular problem: 'The higher the incidence of a recessive gene, the more difficult it is to distinguish from a dominant. Conversely, it is difficult to show that a common gene with incomplete dominance is not really recessive.' As clarifications go, it serves confusingly well in introducing human inheritance. It provides ample warning that ratios of 3 to 1, or 1 to 2 to 1, the classical backbones of genetics, can be well concealed in the hurly-burly of actual pedigrees. Even Mendel's results in his own monastery garden were not quite as clear-cut in practice as he expected from theory. It is now presumed he must have mentioned his expectations to a gardener because the recorded precision of many of his results has never been equalled. They were too good, too much like theory, less like practice, less like life. Life is altogether more complex.

On then to inheritance. At the outset it needs to be emphasized that man is a sexual animal. In any family tree not only are males and females inevitably involved but this basic form of polymorphism frequently leads to a different response by males and females to a similar inheritance. The gene for baldness, for example, behaves as a dominant in men and as a recessive in women; or, to rephrase the same fact, heterozygous women (who have received only one gene for baldness) are not bald while hetero-zygous men (still with only one such gene) are bald. As a consequence of this one sex difference the basic facts about the very basic characteristic of baldness serve as a good opening to the difficulty. Nothing is quite so straightforward as we might wish it to be and here, to help to make this point, is some of the bald information.

Baldness is more common in men than in women (even though equal numbers of both are heterozygotes)

A man may be bald without either of his parents being bald (the gene may have come from his mother)

Half, or slightly more, of a bald man's sons are bald, but only an occasional daughter (the males may have received bald genes from either mother or father, but bald females must have received bald genes from both parents)

All sons of bald females must be bald (as she had a double dose and the sons must receive at least one of them)

A girl is never bald unless her father was bald (both he and her mother had to pass on the gene)

When both parents are bald, all the sons will be bald but not all of the daughters (the girls will have to receive a double dose, and may not have received a bald gene from their fathers)

All of this is based on the assumption that only one pair of genes is in-volved in creating the familiar form of 'pattern baldness', but there is more than one type of hairless head. The ordinary kind is the type that begins early in men and spreads over all the upper surface of the head. There are others that start later and are less wide-ranging in their effects; they confuse the picture even though, as human features go, the inheritance of pattern baldness, the main kind, is relatively clear.

For the converse of baldness, to mention a feature no longer recogniz-able on any hairless head, there is the hair whorl. This spiral at the back of the parting may be clockwise as seen from above (hereafter cw), or anti-clockwise (hereafter ACW). Sometimes there is a double whorl and even, on occasion, three or four, but a study made upon this gyrating property showed that 80% of Europeans were ACW (when it was single) as against 40% of Japanese. In family studies parents who were both ACW had children in the proportion of 5 ACW to 1 CW. If the parents were mixed the children were in the proportion of 2 ACW to 1 CW. Genetically it would appear from all these heads that a single pair of Mendelian genes is operating, with ACW dominant to CW. With double whorls the two gyrations are most likely to be in different directions and no one yet knows what manner of genes are causing them.

Reverting to sex for a moment there are various genes on our sex chromosomes. Consequently there are linkages between the effects of these genes and a person's sex. Probably the most famous example is haemophilia, with the defective gene being on the X chromosome. The linkage explains the fact that no woman is ever (well, hardly ever) the victim of haemophilia, that no man can pass on the defective gene to his sons and that a man can only acquire the disease from his mother. As he is XY the defective X can be most wretchedly effective; as she is XX the sound X is there to mask the defective one. Baldness is entirely different in that it is the masculine environment, and all those masculine hormones, that provides the essential ingredient for the gene to show itself.

Different yet again are various diseases that occur more readily in one or other sex. For example, males suffer more from:

Inguinal hernia	Duodenal and peptic ulcers	Talipes
Phthisis	Kidney-stones	Pyloric stenosis
Various cancers (as of the tongue, mouth, lip, bladder, rectum)		

Females suffer more from:

Migraine	Gall-stones	Femoral hernia
Diabetes mellitus	Mitral stenosis	Splenic anaemia
Myasthenia gravis	Rheumatoid arthritis	
Congenital dislocation of the hip		

These sex differences must relate to the different environment of the two sorts of human being. After all, there are many sex differences, like those involving metabolism, numbers of red cells, hormones, longevity, and more and more, and it is likely that diseases would react differently in both situations. There are also some distinctions between the races, but as yet no human racial character is known to be sex-linked.

The X chromosome is much bigger than the Y of the male. It is therefore entirely understandable that more genes have been discovered to be part of it. Apart from haemophilia their effects include:

Colour-blindness	Night-blindness	Toothlessness
Icthyosis	Myopia	Congenital cataract
Dark-brown teeth	Glaucoma	Addison's disease
Diabetes insipidus	Microphthalmia	Muscular dystrophy
Nystagmus	Parkinsonism	Spastic tetraplegia

Far fewer genes have been discovered on the smaller Y chromosome and even those that have been found – black hairs in the ears, webbed toes, a different form of icthyosis – are much rarer.

While on this subject of linkages there are also a few associations known to exist on the autosomes, the twenty-two chromosome pairs that are not the sex chromosomes. If colour-blindness and sex go together, why should not two other inherent characteristics? The answer is that they do, and there must be linked associations, but they are exceedingly difficult to detect. A parallel in animals is the fact that pure white cats with blue or blue-grey eyes are invariably deaf. One particular gene is responsible both for the genetic error (deafness) and for another characteristic. This other half of the association can therefore act as a warning signal: in short, observe the presence of a particular distinguishing feature and then watch out for the allied disease. Unfortunately the human facts, so far, are not that helpful.

Thirty years ago there seemed to be a multitude of genetically linked associations. Claims were made for: colour-blindness and a crooked little finger; cataract and curly hair; glaucoma and dark eyes; albinism and mental derangement; absence of incisors and dark hair; eye and hair colour with diabetes mellitus; finger-print patterns and blood groups, and more. It looked, as they say, good. Unfortunately the more that people looked the less good it became. By 1969 there were only five autosomal linkages that were generally accepted and a few others that were less certain. Not one was particularly dramatic. They included: the position of the ABO blood-group genes and poorly formed nails, mainly of the thumb and index finger; a congenital cataract and the Duffy blood group; certain aspects of haemoglobin production and thalassaemia disease. It was all a far cry from the well-mapped inheritance of, say, the fly *Drosophila* and, even now, not one of those human linkages has been ascribed to a particular chromosome. Some very recent work, by techniques involving cell fusion in culture, is beginning to crack this problem of linkages being assigned to particular chromosomes. It should therefore not be long before the fruit fly's pre-eminence at last declines in favour of mankind.

The greatest number of human physical peculiarities that can be observed from the outside have to do with the hands and the eyes. The greatest amount of genetic variation so far discovered in any part of the human frame is associated with blood. There are enormous books on the genetics of blood alone, and societies from Tierra del Fuego via

Manhattan to Novaya Zemlya have had blood needled out of them for fresh facts about Duffy, Kell, Lutheran, MNSs and all the others including ABO, the founding father of this particular line of research. However, first to the relative straightforwardness of our digits and to the many varieties of this five-fingered theme.

The earliest example of Mendelian inheritance demonstrated in man was brachydactyly, a dominant gene causing extremely short fingers. There is also minor brachydactyly in which the fingers are slightly longer than in the normal form of this abnormality, save for an extremely short middle phalanx. There is also, as every mother seems to know (despite its extreme rareness relative to scores, if not hundreds, of other quirks of inheritance), polydactyly with the usual form of this dominant gene causing hexadactyly, namely one extra finger or thumb or toe. The Giant of Gath, 2 Samuel XXI:20, had six fingers and six toes. So did a Roman, referred to by Pliny, have six fingers, earning for him the name of Sedigitus. A nobleman of those days had two such daughters, and they too suffered the same spark of Latin originality by being called Sedigitae. There is also another type of hyperdactyly, with a doubling of the end bone of the thumb, and the curious pedunculated postminimi which are small finger-like additions to the hand. Fortunately, these vanish on their own, but most extras have to be severed surgically, not so much for aesthetics but because they interfere with normal dexterity.

To hurry on, and provide amazement that any infant hand is ever normal, there is: lobster claw (where two digits on either the hand or the foot are much larger than normal); ectrodactyly (an absence of digits); talipes (or club-foot, a bunching of the bones that affects about 1 in 1,200 babies); crooked little fingers (what it says); hypophalangy (absence of certain phalanges – each segment of a digit being a phalanx); webbing (almost always between toes II and III); clubbed fingers and toes (more frequent in males); congenital flatfoot; apical dystrophy (different from general brachydactyly, in that fingers are short due to the absence of terminal phalanges); hyperextensibility of the metacarpo-phalangeal joints (the medical abbreviation for double-jointed fingers); Morvan's syndrome (abnormal nails); symphalangism (stiff fingers); brachymesophalangy (short middle joint in fingers and toes); three-jointed thumb; anonychia of thumbs (no nails); and perhaps that is sufficient.

Virtually all of these are dominant genes. It is a reiteration to say so, excusable solely for its importance, but all such traits that run in families are almost bound to be dominant. An abnormal and harmful recessive gene is much harder to spot, whereas a dominant gene will show up wherever it is present. The *Sunday Times* (of London) once showed pairs of pictures of famous antecedents and current descendants. Queen Catherine Howard, for example, born 1521, did look remarkably similar to Lady Jane Howard, born 1945. Robert Devereux, Earl of Essex, born 1566, was similar in appearance to Robert Devereux, Viscount Hereford, born 1932. Between the sixteenth and the twentieth centuries there have been about sixteen generations, and of course any one individual only passes on 50% of his genes to any of his children. With that number of generations this means a steady division of the original genetic complement, adding up to a fraction of one over (about) 66,000. The modern Howards and Devereux, even if on the direct line, could therefore expect to have 0.0015% of their genes in common with their famous ancestors of that earlier Elizabethan age.

How then can there be similarities in appearance? Firstly, everyone is inbred to a considerable extent, not least the aristocracy, thus helping the original gene stock to stay within the family. Secondly, the crucial facial likeness, as with the Hapsburgs and their famous jaw, can be the result of one dominant gene. Half of the children will possess this gene, and so will half of their children. With several chances (or children) in each generation there is a distinct possibility of the gene remaining within the direct line. Should it be lost, and should near-relations then be married into that direct line, there is a possibility of recapturing the gene from those near-relations. It is therefore possible, if unlikely, for Lady Jane Howard to look like her unfortunate and executed forbear because a dominant gene has been crucially involved in the creation of their joint appearance.

With eyes it is colour that has been best examined but, as has already been hinted, even the most important eye rule of them all – blue-eyed parents produce blue-eyed children – has its exceptions. In one Danish survey the results were:

	Blue	Brown	Grey-green or blue-green
Blue father/Blue mother	625	12	7
Blue father/Brown mother	317	322	9
Brown father/Brown mother	25	82	–

The Blue/Blue rule is almost perfect, but not quite, and a brown or green child is not proof, by itself, of anything (such as infidelity). The

Brown/Brown results are as near that Mendelian ratio of 3 to 1 as makes no difference, and the offspring of Brown-Blue marriages are also and correctly as near 50–50 as they should, genetically, be (assuming all these Brown mothers have only one brown gene each). However, there is also a sex factor operating: brown and female tend to go together. Another survey, relating to the children of marriages where the mother had blue eyes and the father had brown, helps to underline this point. (With all such surveys one thinks again of that south London discovery of 30% of the children being provenly sired by an unknown number of other men, and perhaps all offspring should be presumed illegitimate until strong indications of fidelity have been encountered. The thought is a disarming one but is best placed firmly to one side.) Here, therefore, are the results of those Blue mother/Brown father matings:

Eye colour of children	Sons	Daughters
Blue	63	50
Brown	65	81
Grey-green or blue-green	4	2

Note that the sons are virtually 50–50 for blue and brown while the daughters are nothing of the kind. In the contrary kind of marriage, when the mother has brown eyes and the father blue, there is an excess of blue-eyed sons and daughters. Therefore, not only is there one pair of genes, with brown dominant to blue, but there is also a sex-linked gene influencing the picture.

Still more intriguing is heterochromia iridis, or one blue and the other brown, a possession of one or two people in every thousand. It is certainly inherited, and either mimics the parental condition or reverses it. Colour-blindness, on the other hand, is sex-linked. The defective gene is passed by a man to all his daughters, but to none of his sons, and every defective male must have received the gene from his mother. The defect varies considerably between the differing populations of mankind, being most prevalent among those who are most distant in time from having to make their living in a natural fashion. The proportion is about 8% for Europeans, less in general for Asians, and least – perhaps 1 or 2% – for Amer-Indians and Australian aborigines. (There has, incidentally, been much talk about the colour perception of early man, partly because blue and green are unknown in cave art and also because many languages, such as Australian, have an inadequacy of words for the variety of colours. Even Homer's recourse to a wine-dark sea has been imputed as further evidence for primitive colour perception. The controversy has all the raw material for permanent discussion: hardly any facts and no means of proof.)

Like colour-blindness, so does myopia – short-sightedness – seem to accumulate in civilization. One assumes that the selective pressures against it are not so great in unnatural societies, and once again it is necessary to carve through the exceptions to try to find some general ruling. Certainly it appears as if girls are more prone to it than boys. If either a father or a mother is myopic the proportion of similarly defective children is two girls to one boy. If both parents are normal-sighted the children are roughly 3 to 1, normal to myopic. If both parents are defective all their children will be. Nevertheless, there are also exceptions and there is more than one kind of myopia; the rules are not sacrosanct.

As with so much of inheritance, one longs for greater precision, for more rules, for fewer exceptions, and with blood at least a portion of these wishes have been granted.

The blood-group story starts off, clearly and satisfactorily, at the very start of this century. Karl Landsteiner found in 1900 that serum (blood liquid) from some people coagulated the red corpuscles of other people; he coined the initials A, B, O and AB to describe the four groups and, by his work, opened up the blessing of blood transfusions. However, he had also opened up far more than anyone could have predicted at the time and the subject is now enormous. By 1910 the Mendelian inheritance of Landsteiner's groups had been recognized. By 1924 the mathematics involved had been precisely calculated. By 1927 Landsteiner himself and a colleague encountered the MN and the P systems. Unimportant for blood transfusion, in that their mixtures do not cause that lethal clumping of the ABO system, their significance for genetics in general, for anthropology, and most assuredly for immunology has been tremendous.

By 1940 the Rh groups were being added, plus further divisions of the MN and the P. The Rhesus system was the second important medical group after the ABO because it caused 1 in 200 of babies to be severely affected, and frequently to die, due to the blood-group incompatibility of their parents. Now, most fortunately, a way has been devised for preventing Rh-negative mothers from producing antibodies against their Rh-positive offspring. The trickle of groups, medically interesting or not, grew into a flood and were always absorbing immunologically. The Lutheran, Kell and Lewis groups (named after possessors, not discoverers) were all first detected in 1946. Then came Duffy (in 1950), Kidd (in 1951),

Diego (in 1954) – which was found to be virtually confined to people of Mongolian extraction, Yt (in 1956), Auberger (in 1961) – of which only one example has so far been found, she dying that same year, and Dombrock (in 1965). There are others, variants, doubtfuls, inconclusives, or possible future certainties.

'We are all the same,' we say. 'We are all of one blood,' said Mowgli, stretching the net even wider. We are, of course, nothing of the sort. With the major blood groups alone we fall into very separate camps. White men's blood is not the same as black men's blood in its proportioning of all those groups. This particular blood, say the serologists, is probably blood from such and such a population. This blood, say the criminologists, was probably left by that particular man. Fifteen years ago at least 300,000 different blood-group combinations were considered possible; the figure today is very much higher, even if it means less and less as the years, and the discoveries, mount up. Do you know who, or what, fifteen years ago was an O, MSNS, P_1, CDe/cde, Lu^bLu^b, kk, Le^bLe^b, Fy^aFy^b, Jk^aJk^b? It was the commonest kind of Englishman. The most frequent kind today must have an even greater number of letters after his name and be rarer still. Even that particular and commonest combination only occurred in 1 out of every 270 people. 'There is no sign that the pace of developments in knowledge of blood groups is slackening,' says the publisher's introduction to yet another edition of the principal textbook on the subject. Introductory acclamations by publishers are not always at the topmost pinnacle of truth, but in this particular case the remark verges almost on understatement.

By way of concluding the blood topic I wonder if paternity clinics will arise in response to public demand. Many people, particularly those troubled by the division of large sums among various claimants, are obsessed with the rightfulness and general legitimacy of their heirs. A sample of blood cannot lead to proof that a certain child is the product of a certain marriage; it can far more readily prove the converse, if the converse be true. Here, for example, are the harsh facts of life as dictated by the ABO groups (and quoted from R. R. Race and Ruth Sanger):

Matings	Possible children
$O \times O$	O
$O \times A_1$	O, A_1, A_2
$O \times A_2$	O, A_2
$O \times B$	O, B
$O \times A_1B$	A_1, B

Matings	Possible children
$O \times A_2B$	A_2, B
$A_1 \times A_1$	O, A_1, A_2
$A_1 \times A_2$	O, A_1, A_2
$A_1 \times B$	O, A_1, A_2, B, A_1B, A_2B
$A_1 \times A_1B$	A_1, B, A_1B, A_2B
$A_1 \times A_2B$	A_1, A_2, B, A_1B, A_2B
$A_2 \times A_2$	O, A_2
$A_2 \times B$	O, A_2, B, A_2B
$A_2 \times A_1B$	A_1, B, A_2B
$A_2 \times A_2B$	A_2, B, A_2B
$B \times B$	O, B
$B \times A_1B$	A_1, B, A_1B
$B \times A_2B$	A_2, B, A_2B
$A_1B \times A_1B$	A_1, B, A_1B
$A_1B \times A_2B$	A_1, B, A_1B, A_2B
$A_2B \times A_2B$	A_2, B, A_2B

And that, all of it, only relates to one blood group's particular set of rules. It will free about 20% of males if they are wrongfully accused of fathering a child. It, and all the other groups, will help to provide the necessary peace of mind for a family and its inheritance or, contrariwise, do nothing of the sort.

There are certain classic examples of inheritance that are irrelevant to most of us but are striking, all the same. There is albinism, possessed by 1 in 280 people as a recessive gene, however ignorant they may be of the fact. It is inconsequential unless one happens to mate with another who also carries it. The result is that only 1 in 20,000 offspring is affected just as Noah – or so it is alleged – was affected. There is also total deafness, which in the past was immediately coupled, and should never have been, with mutism. It was the first trait to be studied from a eugenical point of view with the recommendation, recessive genes being involved, that the deaf should not marry each other. When they do so the consequence is a considerable number of deaf children. One Ohio survey of thirty-one double-deaf parents revealed that 70% of their eighty-nine children were also deaf. Unfortunately such deaf partnerships, whatever the defects of creating more deaf children, tend to be happier for the parents and, if the custom prevails, will ensure the continuance and even the increase of inherited deafness. R. Ruggles Gates reported in 1946 that the School for the Deaf in Pennsylvania was already educating the grandchildren of some of its former pupils. (He also went on to say that sterilization would there-fore be desirable in some cases.)

Naevi, or moles, have been the subject of considerable familial attention around the newborn bedside. Undoubtedly there have been similar such birth-marks occurring from one generation to the next, in which case a dominant gene must be at work, but most people's naevoid peculiarities seem to be their own in that they bear no links with those of their kin. A group of identical and dissimilar twins were examined for their marks; the average of 18.5 per person varied in position and number but showed greater similarity among the identical than among the dissimilar twins. Certainly inheritance is involved but the rules are, as scientists tend to say when near-total ignorance exists, incompletely understood.

We all have hair, but none of us has any on the terminal segments on any finger. Lots of us have hair on the first segment but it is the middle segment that is the most interesting. A generalization is that no child has hairier middle segments, either in number of hairs or in number of digits involved, than the hairier of its two parents. Children of both sexes are equally finger-hairy below the age of eighteen, but thereafter males grow more, just as they do with eyebrows. Irish are said to have less mid-digital hair than other North Europeans, Italians have still less and, going further south, Negroes have least of all.

Hypertrichosis universalis, or dog-face, is extremely rare. Those afflic-ted have their bodies and faces covered with long, silky, yellowish hair that starts to grow in childhood. Hypertrichosis of a kind, certainly less absolute but extremely marked nonetheless, is fairly common, varying upwards from a normal hairiness of the chest, shoulders, back and legs, to an extreme woolliness that is very striking. The genetics of all this, fascinating as it would be, is quite unknown. Most certainly there is racial inheritance linked with hair, for the Caucasoids have most, the Mongol-oids less, and the Negroids least of all. Complete hairlessness, or alopecia congenita, can be produced by disease or it can be inherited and, if in-herited, it is either dominant, recessive, or sex-linked. The fact must be becoming disturbingly plain that only rarely can anything be called straightforward in human genetics.

It may be as well to interject a further point about mutations. They have occurred in the past, both to make us different from everything else and to create the inherited differences among us, but they are also occurring today. Any gene present in a child that was not present in his parents must have arisen by mutation. It is possible, therefore, for a child to have the visible effect of some dominant gene, say the one causing brown eyes, without either of his parents possessing that gene or its effect. Mutations do not have to exert some change that has never been known before; in fact they are much more likely to occur where successful mutations have happened in the past. Therefore haemophilia, for example, can suddenly arise in a family which has previously been quite free of the disease. Not one of Queen Victoria's antecedents had ever been plagued by this clotting defect; the mutation must therefore have arisen in her, her father or her mother (although her mother had already had two apparently normal children by an earlier marriage).

It is possible to calculate the rate at which a particular mutation is recurring. E. T. Mørch, of Denmark, working on the incidence of chondrodystrophic dwarfism, produced a classic paper on the subject in 1941. This disease is inherited as a single dominant, with almost '100% penetration' – the genetical way of saying that it actually manifests itself (almost) always. At one Danish hospital, during a period of thirty years, 94,073 babies were born of which ten were afflicted with this cartilage disorder that leads to abnormally short legs and arms. Only two of these dwarfs had a parent with the same condition; therefore the other eight must have arisen by mutation. With two genes per baby that could have mutated to give the condition, and with those eight particular babies, the actual mutation rate can therefore be calculated. If 8 is divided by 188,146 (or two genes each for 94,073 babies) the answer is $4,252 \times 10^{-5}$, or one mutant of this kind in every 23,518 genes.

As mutation rates go, this is high. Muscular dystrophy is thought to arise spontaneously at a similar rate, but most others for which rates have been calculated, such as haemophilia, deaf mutism and Huntington's chorea, are all considerably lower. However much advice genetic counsellors may provide, however much they may succeed in discouraging matings that will perpetuate harmful genes, their work will always be counterbalanced to some extent by the spontaneous arrival of new mutations. Haemophilia, muscular dystrophy and the rest will always be with us, or rather with that small fraction of us who will suffer them, whether from inheritance or recent mutation. So far mankind has only found a way of inducing mutations, and has not – as yet – the glimmer of an idea for a method of reducing them.

The traditional cot-side story about ears is that adherent ear lobes are a sign of wickedness or worse, but genetics, alas, can find no such correlation. Women have them more than men, roughly in the ratio of 2 to 1. When both parents have adherent ear lobes no child of theirs will have lobes, or so it has been reported, less than three-fifths adherent. Conversely, when both parents have lobes less than three-fifths adherent, no child will have fully adherent lobes. Only a minority of Europeans have their lobes attached while Negroes are customarily lobe-less. Not only is there difficulty in making conclusions about ear genetics, but even agreed descriptions of ear shapes, and whether or not they have tubercles, pits, cups, ridges, auricles or any other oddnesses of that external pinna, have not been achieved. Examine two ears yourself and then attempt to describe them, let alone classify them. Also, while at this task, spare a thought for the stickiness or dryness of wax. Mongoloids in general have dry, odourless ear-wax, and if Japanese do have sticky wax it has been inherited as a dominant gene. Negroes, however, are 100% sticky, Europeans 70%, Chinese 3%. (B. Adachi, a Japanese, seems to have done most of the work on this score in 1937.)

Both ear secretion and arm-pit secretion are connected with sweat glands, and these are more numerous among people with arm-pit odour. Perhaps this too is a subject that has lacked the spotlight of merited attention. An early inquirer was Richard Burton, explorer and considerable investigator of mankind. He was interested, for example, in the differences between West African Negroes and various Mohammedanized tribes. 'The earliest distinction between them consisted not in external features but in smell,' he wrote. 'That was the best test, and the difference was occasioned by a different development of the sebaceous glands.' The destruction of so much of Burton's material by his widow is a loss insufficiently lamented.

Twinning can be inherited. Certain families seem to be particularly prone to it and certain populations are more prone. These tendencies apply only to two-egg (dissimilar) twins and not to one-egg (identical) twins. Negroes, for example, produce a lot, Europeans rather fewer and various groups of Mongoloids fewer still. In general, the sister of a woman who has borne two-egg twins stands about double the normal chance of producing twins herself. As the rate in Europe and North America varies between one birth in 70 and one birth in 125 the chances for a sibling sister are therefore somewhere between 1 in 35 and 1 in 62: the event is still a rarity. It is less relevant whether a brother has already fathered twins; the inheritance comes mainly through the female line.

Perhaps more will be discovered about the strange phenomenon of twinning when the recent dramatic change in its rate is understood. In Britain in the late 1950s the number of two-egg twins suddenly began to drop. Since then the fall has been from 9.2 per thousand births in England and Wales to 7, and in Scotland from 10.2 per thousand to 7.2. Similar falls have happened in Australia, Belgium, Denmark, Holland, Italy, New Zealand, Sweden and Switzerland. Lesser falls have been observed in Portugal, Spain and Japan. The odd one out is the United States where the number of two-egg twins during the same period has remained similar. It is difficult to produce a theory that explains the general drop *and* America's exclusion. For example, older mothers produce more twins and mothers have been becoming younger, but so they have in America. Hormones readily capable of affecting humans have been used in fattening cattle and poultry, but so they have in America. (The alleged villain of the piece, diethylstilboestrol, is now forbidden in the United States, but its banning only took effect in August 1972.) One thing is certain; some change in the environment and not in the genetics must have been responsible for such a rapid shift.

The chief bonus of twins has been the assistance they have given to that constant question: has a feature been caused environmentally or genetically? Was it nurture or nature? One-egg twins reared apart have the same genes but different homes, and two-egg twins reared together have different genes (although 50% of them are shared, as with all siblings) in the same home. Many of the facts embedded within this book have originated from such comparisons. Twins complicate pregnancies, they die more frequently than singletons, they cause their mothers to die more often in childbirth than do normal deliveries, and they can be an unexpected blow to any family budget, but geneticists and others feel considerable gratitude at this further expression of human abnormality.

The initial intention in this chapter was to separate normal inheritance from the inheritance of disease, but the resolve has failed in the preceding pages. Diseases are so useful; they help to explain so much. Consequently many have already been mentioned (such as haemophilia, dystrophy, chorea), but the forthcoming pages deal exclusively with disease and with the particularly telling examples. In their turn they should help prepare the ground for further subject matter most forcibly linked with inheritance, such as genetic counselling, prevailing ethics, eugenics and evolution. The situation today is one of imminent change, and the inheritance of disease has done much to unsettle many of the old traditions, the old morals. Disease is important on its own account as it cuts swathes through human happiness; it will be doubly important should it make similar inroads upon human custom.

That number of 1,500 distinguishable diseases known to be genetically determined in man will certainly increase with increasing knowledge. These disorders are either chromosomal (as with mongolism) or genic in that a faulty gene or complex of genes is involved. It was calculated, for example in one North American paediatric hospital, that 30% of its admissions were due to genetic disorder. In Britain it has been reckoned that 40% of all child deaths are more or less related to genetic disease. Much mental illness is genetically based, and half the hospital beds in the world are occupied by the mentally sick. About 70% of childhood deafness, and 80% of blindness in non-tropical countries is caused genetically. A total of fifteen million Americans, according to the National Institutes of Health, are thought to be affected by genetic disease.

Not every congenital malformation is caused by genetic disorder, and it is necessary to distinguish the two of them. A genetic disorder is one that has been inherited or has been caused by a recent genetic mutation. A congenital error is one present at birth, and not necessarily genetically caused, such as the short limbs of the thalidomide victims, but some congenital *and* genetical errors – confusingly – may not actually be detectable on that first day or even during the first year. Huntington's chorea is an extreme example which does not manifest its lethal presence until middle age. It and many other abnormalities are known to be caused genetically, but no one knows how many, partly because human genetics is being unco-operative in yielding up facts and also for the reason that environmental influences within the womb are proving equally obscure. (Thalidomide did at least have the blessing of causing a rare malformation. Had it produced one of the commoner variety its malevolence would have remained undetected for much longer, possibly for ever.)

Nevertheless, the congenital malformations serve to underline, whatever the exact proportion may be, the importance of 'the inherited errors of metabolism', as a famous early paper first referred to genetic disorder. There are plenty of them. Here are some figures (taken from one area of Quebec) of fatal malformations, all of which were congenital and all of which had their lethal effect before the affected infants were one year old. The systems involved were:

Circulatory	220	deaths per 100,000 live births				
Nervous	210	„	„	„	„	„
Digestive	60	„	„	„	„	„
Genito-urinary	40	„	„	„	„	„
Musculo-skeletal	30	„	„	„	„	„

These add up to the considerable total of 560 deaths for every 100,000 live births. Congenital malformations are in fact even more damaging in that many stillbirths are directly attributable to them, so too many deaths after the age of one. The numbers are formidable, but have achieved great prominence only since infectious diseases have been demoted as the principal cause of infant death. England and Wales are no more unfortunate than any other area, but 14,407 babies were born there in 1971 with congenital malformations of one kind or another. Mongolism, or more correctly Down's syndrome (named after the nineteenth-century English doctor who made the classical description), is often spoken of as if it is the only congenital abnormality. But approximately 2% of all live births possess a severe defect while only 0.06% of them are linked with mongolism.

It is as well to remember, if the figure of 2% abnormal indicates perfection for 98% of us, that the truth is nothing of the kind. We too are largely faulty, whether we show it or not. W. F. Bodmer, of Oxford, estimates that virtually everyone carries abnormal genes. Therefore, if a determined attempt is made to clean up the human race genetically, to rid it of deleterious genes, only a minute number of us would be allowed to breed. In short, we are almost all a potential threat to succeeding generations, but happen to have the good fortune not to be afflicted ourselves. Our genes are at fault even if we do not show the fact or suffer from it or even know about it.

Congenital malformations, whether caused genetically or not, vary around the world. Anencephaly, for example, the condition of brainlessness, has a frequency of 400 per 100,000 births in Northern Ireland, 200 in England, 80 in Japan and 50 in West Africa. Cleft lip, with or without the cleft palate often linked with it, varies from 300 per 100,000 in Japan to 100 in England, Denmark and North America and to 40 in Nigeria. Talipes, or club-foot, affects 400 in 100,000 of all Maori live births but 100 almost everywhere else. Mongolism varies more with maternal age, affecting 140 out of 100,000 in general but 1,000 out of 100,000 if the mother is over forty.

The form of amaurotic idiocy known as Tay-Sachs disease, mentioned at the start of this book for the dilemma that it presents notably to American Jews from eastern Europe, was originally identified by W. Tay and B. Sachs in the last century when they gave an account of the greatly swollen neurons in the brain. It was the first of various genetically controlled idiocies to be described, with paralysis and blindness following the initial symptoms. At present there is no cure or even a treatment for Tay-Sachs, which kills its victims between the ages of 6 months and 3–4 years, but at least much of the misery involved can be side-tracked. By taking a sample of amniotic fluid, and by analysing the cells for the presence, or absence, of the crucial enzyme hexosaminidase-A, it is possible to know if the child is a victim. If that essential enzyme is missing there is time for an abortion, if the mother so desires on hearing the news. It is also possible, by the donation of a blood sample, for a person to be informed if he or she is a carrier of the recessive gene that causes the trouble. If both partners of a prospective liaison do have the gene they should be warned of the problems if they decide to go ahead with the marriage, have pregnancies and each time suffer the narrow odds of 3 to 1 against Tay-Sachs.

The number of hereditary diseases that can be diagnosed in the uterus before birth is increasing all the time and will continue to do so. As of now the complete list, presented recently to the World Health Organization, is considerable (see Appendix 2). Early diagnosis of such diseases by the examination of a sample of amniotic fluid will undoubtedly become a commoner procedure in the future, and until such time as a better system is devised. Against the prevailing medical ethic is the demand by many mothers that a similar investigation should be carried out to detect the sex of their unborn child. Current amniocentesis techniques are not without risk; therefore the prevailing opinion is that the method should only be used when there is a strong indication that some hereditary disease might be involved.

The greatest controversy of all concerning congenital malformations has raged around the commonest defect to affect the central nervous system. Both meningocele and meningomyelocele are lumped together as spina bifida, and both involve a failure of the developing spinal canal to close completely. The error can vary in its severity, but it leads more often than not to death. If no treatment is given a victim of spina bifida is usually dead within a few weeks of birth, probably from infection.

Doctors have traditionally dedicated themselves to two causes; the prevention of suffering and the preservation of life. These aims can be incompatible, notably at the end of life. They can also provide a cruel paradox, as with spina bifida, where improvements in medical technology have meant that more and more children are growing up with severe physical handicap. In the late 1950s and early 1960s a modern form of surgical treatment for bifida was introduced. If performed very promptly, and on the first day of life, many more of the victims could be made to survive. The new policy was energetically pursued, the mortality fell below 50% and it was generally presumed that yet another medical advance was being satisfactorily accomplished. Then came doubts.

Is life the criterion, or is quality of life? Under the new regime the infants had to experience not just one operation on the first traumatic day, but many more during the subsequent years. About half of these children lived long enough to encounter school. Of this 50% about four-fifths were mentally handicapped, with an IQ of less than 80, and all were physically handicapped. Most were paralysed from the waist down, which meant incontinence and an inability to walk. Wheelchair existence led to poor bones, repeated fractures and dislocations, and some of the victims were blind. Was this the quality of life made possible by another advance of surgery?

The degree of intelligence often helps to provide a guide line, as with that mongol case at the beginning of this book, but it does not help with spina bifida. A doctor has written: 'I question the assumption that severe physical handicap (life in a wheelchair, faecal incontinence, urinary diversion, and impotence accompanied by normal libido) is more tolerable if the intelligence is preserved than if it is not.'

When surgery is not carried out the victim does not die immediately. One consultant paediatric surgeon, in favour of abandoning much of today's protracted diligence, raised further bifida problems: 'In my opinion it is quite impossible to kill off such a baby, but if surgical treatment is withheld then it is only reasonable to withhold other forms of treatment such as antibiotics, oxygen and tube feeding.... If a baby is not to be treated then the surgeons and nursing staff should do nothing to prolong life.' In which case the medical paradox is entirely removed; neither the short-term prevention of suffering nor the long-term preservation of life is being pursued.

An incident in Hull provided further ethical confusion. A child born in this Yorkshire town had spina bifida complicated by hydrocephalus, as is often – 85 % – the case. Medical opinion held that surgery was necessary to save the boy's life, but the parents refused to give their permission. 'We love our son like any other parents, but we don't want him to lie in hospital suffering operation after operation and be used as a guinea pig.' The director of social services, acting under the Hull Social Services Committee, therefore applied to the court for the child to be taken into the care of the local authority. The committee felt it had 'a clear duty to have regard to the welfare of the child in its widest sense and is satisfied ... that it will be in the child's best interests for him to undergo this operation'. The court made the required order and the committee, after further deliberation, gave its consent to the operation. The parents said: 'It was heart-breaking enough to learn that our son was handicapped, without doctors prolonging our agony. This dreadful dilemma could befall any family at any time, and they might find themselves getting the rubber-stamp treatment.' It is easy to imagine yet greater vehemence coming from different parents overruled in such a fashion.

On the other hand there is Angela Dowson, now aged 29, pretty and quite independent. 'I know my mother thought I'd be better off dead when I was born with spina bifida. And all these articles you read say how terrible it is to let such babies live. ... No one ever tells you how little it matters to have spina bifida and a wooden leg.' Plainly a most exceptional person. Her leg had been amputated after a cut had become septic.

However, no one is suggesting that all victims of this one congenital malformation are similarly afflicted, and the need for a more selective approach is now being advocated. A major follow-up survey in the Sheffield area of England provided some essential facts. The policy there between 1962 and 1964, when 200 severely afflicted bifida children were born, was that every one of them should receive its first surgery on the very first day of its life. At the time of these prompt operations 151 babies were classified as 'severely afflicted' and 49 less so.

Of the 151 only 62 were alive by 1971. Of these 62 none could walk without crutches or callipers and only 5 had an IQ over 100. Of the 49 who had been less severely affected at birth only 8 had since died. Of the remainder – 41 – all were physically handicapped but 8 had an IQ over 100 and 17 had an IQ between 80 and 99. To lose a total of 97 out of the original 200 implies considerable distress, despite and because of the surgical efforts involved, but at least the death rate was only 16% in the better group as against 59% in the severe group. The 103 still alive eight years after the trials began were either fully or less acutely aware of their predicament. 'What right,' wrote Geoffrey Hatcher, 'have I to recommend withholding treatment? But what right have I to inflict major chronic suffering on another human being and on his family?'

Putting the size of the problem into perspective, many more spina bifida victims survive every year in Britain than were affected in Britain by the drug thalidomide. No one knows why the malformation is so common. Also no one knows why it is more prevalent among the poorer sections of the community and among certain populations, such as the Irish.

A hypothesis to account for both facts was put forward in 1972 by J. H. Renwick, of the London School of Medicine and Tropical Hygiene, when he indicted potato blight as a possible cause if eaten or even handled in early pregnancy. Most certainly Ireland grows a lot of potatoes, has severe blight *and* the highest recorded incidence both of spina bifida and anencephaly in the world, but not every country follows suit. Taiwan, for example, consumes hardly any potatoes but also has a high incidence of neural tube defects. The British Department of Health first stated that the Renwick evidence was unconvincing, but within months had changed its tune. Keith Joseph, Social Services Secretary, said in late 1972 that blemished potatoes might have a harmful effect if eaten by an expectant mother in early pregnancy. Perhaps they do, perhaps not. No one knows as yet, but every woman with child can now add potatoes to the list of 1,500 other agents already thought to be teratogenic, to be capable of damaging the small object growing so fearfully within them.

Genetics consistently raises issues along with its facts. Mongolism is no exception and is the predominant single cause of severe mental subnormality. To some extent its prevalence is being reduced because there are fewer older mothers these days, but, as counter-weight, better techniques for keeping more mongols alive for a longer time are being introduced. About 30% of all severely retarded children in the United States and western Europe are afflicted with Down's syndrome and, plainly, it would be beneficial if this proportion could be lessened.

Plans have been proposed. All pregnant women aged forty or more, the highest risk group, could be asked to submit to prenatal screening. The screening would detect a high number of mongols, about 1 in 50 of all foetuses, but older women also suffer the risk of greater complications during abortions should they elect to have one. Even the process of amniocentesis, necessary in the screening programme, might itself lead to abortion and older women are less able than others to conceive again, to make good the loss. Women over forty produce a disproportionately large share of all the mongol children, but they only produce a small share (1.5% in New York) of the baby total; most mongols (84%) are produced by women under forty. If the age-limit for screening is dropped to thirty-five it would be necessary (again for New York) to examine 7% of all pregnancies. Assuming every Down's diagnosis led to an abortion, a total of 35% of the cases would be prevented. The rate of incidence would therefore be reduced from 1.05 per 1,000 live births to 0.68 per 1,000. But would every woman submit to screening? And would every woman, when told that she was carrying a mongol, wish to expel it? Or would they wish that this further advance, this extra blessing of pre-natal detection, had never surfaced to destroy the relative tranquillity of not knowing?

Epileptic women, already with one disability, suffer the extra disadvantage of producing among their children twice as many severe congenital malformations as is normal. In one survey covering 427 pregnancies of 186 epileptic women the commonest anomalies were congenital heart disease, cleft lip and microcephaly. Mental subnormality occurred in 1.5% of the children as against 0.2% in a control group. So what can be done, without anyone reaching for a sterilizer, that is morally acceptable?

The depressing figures may not result solely from bad genes. Perhaps epileptics, suffering from what is still regarded as a social stigma, are more limited in their choice of a mate, with each available partner having a relatively poor genetic background. Or perhaps the epilepsy itself has an effect upon foetal growth, with the transient anoxia associated with a fit being the cause. Or perhaps the very drugs given to help them, the anti-convulsants, are to blame for some of the foetal damage. To produce twice the number of congenital malformations sounds alarming, and indeed early eugenicists advocated that epileptics should be forbidden from child-bearing, but that doubling only raises the proportion from 2% to 4%. Should this be cause for denying epileptic mothers the right to reproduce, to create children of whom 96% are normal instead of 98% as with the rest of us?

Early eugenicists were also fairly severe in their recommendations for diabetics, even though the genetics of this disease is still poorly understood. With 400,000 cases in Britain, 2,000,000 in America and with thousands of undiagnosed victims of diabetes mellitus in every country (suggestions have been made that 1 % of the population even in medically advanced communities does not know it has the disease), considerable numbers of people are involved.

Twins have again been helpful in providing information. In one study of ninety-six identical-twin pairs affected by the disease sixty-five were found to be both diabetic but the other thirty-one were discordant – only one twin was affected. Significantly, if the disease developed before the age of forty, half the pairs were discordant; but, if it developed after that age, almost all pairs were concordant. The fact is intriguing: does it mean that non-genetic factors are more important than had originally been supposed? After all, diabetes is most prevalent in the most affluent societies, and environment, therefore, seems to be playing a part. Or is it merely permitting a genetic proneness to the disease to manifest itself, a proneness stimulated by life in a wealthier community?

An extra fact is that a family history of diabetes is common in concordant twins and largely absent in discordant twins. Are the concordant similarities therefore being ordained by inherited genes, it being reasonable to expect total conformity in identical twins, while the discordant differences are caused by the environment? Unhelpfully diabetes develops late in life, at a time when pairs of identical twins are usually living separate lives and experiencing different environments. Most certainly the problem is more complex now than when those enthusiasts of the new eugenics advocated a policy of non-marriage for all diabetics. On the other hand, the use of insulin (since 1922) has kept hundreds of thousands of diabetics alive who would otherwise have died. The genes that cause, or help to cause, or increase the proneness to diabetes are increasing in the population at an unprecedented rate. The numbers of future diabetics will multiply accordingly.

It is convenient to end this inheritance section with a disease that has its roots firmly in the past and its moral issues equally firmly in the present. Of course all inheritance is the past's contribution to today, and what we do now is our provision for the future, but the story of sickle-cell anaemia strides most purposefully from one to the other. It was James B. Herrick who, in Chicago in 1905, first specifically mentioned the 'sickle-shaped red corpuscles' in a blood film of an anaemic man. This patient was a Negro, and the disease, initially called 'Herrick's anaemia', was a link straight to West Africa and the needs of quite a different environment.

For survival in West Africa it is beneficial to be able to withstand the mosquito-borne attacks of *Plasmodium falciparum*, the parasite responsible for one form of malaria. A mutation in the human haemoglobin molecule, which involved the substitution of the amino-acid valine for glutamic acid, proved to have a curious blessing in malarial West Africa. For some reason the resulting deformity to haemoglobin makes it less susceptible to invasion by the parasite. This same deformity makes the corpuscle less saucer-shaped than normal, more like a sickle, and the current name for the disease has existed since 1922.

By no means is it a blessing, malaria apart, to have sickle-shaped cells. They can block small blood vessels, thereby reducing oxygen supply to the area. They can lead to tissue destruction, anaemia, thrombosis and death. These extreme defects are the evolutionary price that has to be paid for the improved ability of a population to resist malaria. Once again individuals have to suffer for the sake of the community, and the sickling mutation achieves this end by operating as a recessive gene.

Any individual has three possibilities: either two normal genes, or one normal and one sickle, or two sickle genes – a double dose. This last event is likely to lead to death, due to the gross amount of abnormal haemoglobin, and at least 80,000 a year in West Africa are thought to die from this cause. The double dose of normality is good for haemoglobin, but bad for resistance to *Plasmodium*, and most malarial victims are in this category – thousands still die from the parasite wherever it is endemic. Only the mixed category achieve the best of both worlds, a reasonable supply of good haemoglobin and a certain resistance, via the cells that are sickled, to the mosquito's parasitic protozoa.

No portion of the community is therefore left unscathed by this arrangement. The double-normals suffer from malaria; the double-sickles probably die from anaemia; the half-and-halfs, although the best of the three, have much faulty blood. If the population is then transported, as happened for 250 years, to a world where malaria is no longer such an executioner, the conflicting priorities become very different. Natural selection for Negroes in North America has subsequently operated to reduce the amount of sickling. From an estimated 30% carrying the gene in West Africa only 9% of Negro-Americans now do so, the difference having been caused by the lack of an advantageous counter-weight to the disadvantage of sickled cells. The homozygotes, the double-dosages, die out steadily and, with no malaria to speak of, there has been no evolutionary virtue in maintaining the heterozygotes.

How then to be rid of the disease? The straight attack is being actively pursued, with cyanates and even aspirin to increase the uptake of oxygen by the faulty haemoglobin. A more genetic attack has also been tried, not without difficulty. There is a law in Massachusetts stating, in part, that: 'Every child, which the commissioner of public health, by rule or regulation, may determine is susceptible to the disease known as sickle-cell trait or sickle-cell anaemia, shall be required to have a blood test . . .' To supplement the law a procedure has been developed which permits mass, cheap screening of blood samples at 120 an hour.

It would seem entirely beneficial, at least at the outset, to find out who has the disease, to treat the victims if possible, to acquaint them with their inheritance, to advise all sickle heterozygotes of the risks of mating with other heterozygotes, and to inform them that every such double marriage will give the disease to one quarter of the children. The cost of a few cents per sample, enabling the screeners to distinguish between homozygotes (if still alive), heterozygotes and double-normals, would seem to be money well spent. Of every 400 people within America having some West African blood in their veins, one has the disease, forty are carriers and the remainder are free of this aspect of their inheritance. Who, in short, are the forty?

Unfortunately many problems ensue: loss of employment because of sickle-cell detection (following knowledge of this fact by employers); wrongful assumption that some other complaint (appendicitis, heart trouble) is a manifestation of the sickle-cell defect; poor adoption prospects for all proven carriers. *Plasmodium falciparum* flourishes far from Chicago, from Harlem, from Washington DC but is still exerting malevolence all those miles away. The evolutionary expediency, developed by mutation in the humid forests around the Bight of Benin, has been overtaken by events. It is no longer any kind of blessing. It is an error, and it will not go away.

So too for all of us. We are the result of similar expediency in the past. We are not perfectly fashioned after some excellent image both ideal and without imperfection. Within each of us is a backlog of accumulated mutations, some still good, some bad, some irrelevant, some inextricably partnered with others. Two per cent of our offspring have gross and visible defects. These suggest to the rest of us, those lucky to escape their afflictions, that we are therefore whole, well made and without defect. We should revel in our fortune but this should not blind us to the chance contrivance of all our origins and to the faulty genes that we unwittingly pass on to our descendants.

CHAPTER 7

The science of probability

Cornerstone of science – Counselling changes – Average, median, mode, mean – False facts – Quetelet and the normal curve – Height prediction – Group differences – Positive correlation

'When you can, count'

FRANCIS GALTON

'Round numbers are always false'

SAMUEL JOHNSON

'When you can measure what you are speaking about . . . you know something about it; but when you cannot express it in numbers . . . you have scarcely advanced to the stage of Science'

LORD KELVIN

'There are 3 kinds of lies: lies, damned lies, and statistics'

BENJAMIN DISRAELI

'Statistical thinking will one day be as necessary for efficient citizenship as the ability to read and write'

H. G. WELLS

'Probability,' said Mark Kac, 'is a cornerstone of all the sciences, and its daughter, the science of statistics, enters into all human activities.' Consequently a certain digression into the crucial realm of statistics is necessary if the probabilities involved in inheritance are to be encountered with equanimity. Statisticians are often in despair about our appreciation of their work. They are amazed that advertisers can show graphs – booming sales, rising exports – without the mention of a co-ordinate. They worry at our acceptance of some modest sample as the basis for a startling generalization. And they wonder why the 'significance' and 'probability integrals' of biological argument have not become part of everyday vocabulary. Why, in short, is there such a general disregard for statistics, and even a considerable misappropriation of the word by others as a synonym for facts?

In science statements are made, bearing probability in mind, as if they are inviolable laws. The rest of us soon forget about the chance involved, and distrust science the moment we encounter an exception to one of its qualified generalizations. Similarly, when a scientist says that this group is taller than that group, he is immediately liable to misinterpretation by the rest of us. He has not said that every member of this group is taller than every member of that group. He has only stated, in effect, that one normal curve cannot be superimposed precisely upon another normal curve. One group has the edge, however slightly, over the other.

If everyone was more statistically inclined the following question, for example, would receive answers nearer the mark than is generally the case. How many people do there have to be in a room before the chances are better than 50/50 that two of them were born on the same day of the year? There is scope for winning money here as the answer, surprising to most, is twenty-three. By the time there are thirty-one people in the room the odds are 3 to 1 in favour of such a duplication, and the odds rise to 8 to 1 when there are a mere thirty-nine people. Try it at the next suitable gathering. Win money or, possibly the greater gain, fail to do so and learn for all time that probability suggests likelihood but never certainty.

Take even the spinning of a coin which, forgetting about the minor sexual imbalance of the birth ratio, is exactly the same as the boy-girl possibility. The probability of a head (or a boy) at the first toss is 1 in 2, or 50/50. At the second toss, or indeed at any subsequent try, the probability is identical. With most of us, or so the statisticians suspect, there is a feeling that the throwing of a few heads in a row (or a string of boys) steps up the chances – whatever that means – of producing a tail (or a girl) at the next throw. It does not, but the probability of getting two or more heads (or boys) on successive tries is the product of each probability, namely 1 in 2 multiplied by 1 in 2, or 1 in 4. The likelihood, therefore, for three heads (or boys) in succession in a family is 1 in 8, or 7 to 1 against. For five heads (or tails, or sons, or daughters) the probability is 1 in 32, or 31 to 1 against. Therefore, any coin-tosser (or parent) can expect to see five heads together and five tails (or one set of five boys and one set of five girls) in every thirty-two tossings of five coins (or in every 32 five-sib families). It all sounds very reasonable, but not to the parents of any large family of same-sex children. What, they ask their doctor, is wrong with them? What is wrong, he should reply, with your appreciation of probability theory?

Everyone can be at fault about the operation of chance, not just parents and not just gamblers convinced that each run of reds on the roulette wheel increases the odds for a run of blacks. During World War II, and in Bomber Command of the Royal Air Force, there was similar certainty. Approximately 5% of bombers failed to return, or one in twenty for every mission. Therefore the belief grew that each newcomer had a good chance of survival until he began to approach his twentieth mission. The odds against him, it was felt, were increasingly loaded with every mission and, should he actually survive the critical twentieth flight, the man was thereafter living on 'borrowed time'. In fact, remembering the bomber technique of having to fly straight through a well-defined enemy box of anti-aircraft fire, the chances for each bomber occupant remained similar for each flight, irrespective of the number he had already flown. Of course other factors were involved, such as inexperience and fatigue, but the chance of 20 to 1 against being shot down incurred considerable and quite unreasonable deductions. In the realm of probability there is no such thing as borrowed time. As A. J. Ayer put it, 'there are no such things as the laws of chance in the sense in which a law dictates some pattern of events'.

The example of a common birthday and the random fortunes of war are more complex than mere heads and tails, but genetic counselling is yet more so. Perhaps those who are advised that the chances of their producing a malformed baby are, for example, 13 in 147 should immerse themselves in the reality of this statement by examining other probabilities in more familiar contexts. Every poker player should know that the likelihood of being dealt one pair is 1,760 in 4,165 (or slightly worse than 1 in 2). For two pairs the odds are 198 in 4,165. For a full house they are 6 in 4,165 (or 293 times less likely than a pair), while the odds against four of a kind are much more severe, being 1 in 4,165.

Not everyone plays poker, but, as a further example, everyone can have difficulty extracting a pair of socks from the drawer. If ten socks are within and four are extracted the chances will only be 99 in 323 that a pair has been selected. Therefore, in all probability, try again. To be told, by a counsellor, that a particular genetic hazard will arise following similar odds may make most sense if thought of in terms of poker or socks. To play with dice is to be further informed of the vagaries of chance and perhaps every counsellor should have a set. So too those receiving his advice.

There is also the problem of median, mean, mode, normal and average, words that are used most precisely by scientists. Average (or mean) tends to be used in ordinary life where median or mode might be either more profitable or more accurate. The word average, having its origins in the world of insurance and compensation, originally had a limited meaning. When some cargo was lost at sea and some safely delivered the unfortunate losers were compensated in part by those whose goods had arrived. The sea damage was called havaria, and the compensation soon acquired the same name. As money was paid in proportion both to the value lost and to the value received the necessary computing of the 'average' figure laid the foundation for today's use of this much-used word; but it now covers too wide a range.

If ten men are earning £1,000 a year and one other man is earning £10,000 it is misleading, however accurate, to say that average earnings for these eleven men are £1,818 a year. Similarly it is wrong to say, however true, that human beings possess an average of less than two eyes, less than two ears, and less than two feet. It is preferable to use the mode, the number most frequently met with in any series, namely £1,000 a year for those eleven men, and two eyes, ears and feet for us. There is also the median, when 50% of individuals, such as income-earners, are on one side of a central line and 50% are on the other. Finally, there is the range. Of what good is it to a hatter to be told of the average hat size? It is slightly better to be informed of the modal size, better still to be told of the median size as well, and best of all to be informed of the range that goes with these other two.

The current yearning for averages, and planning for the average man, has managed to turn all of us into misfits. It is constantly repeated that the average family has, say, 2.4 children, as if this were either feasible or desirable. None of us has that number of children and, however correct the average, this figure indicates little about typical family size. The American Public Health Association, according to Darrell Huff in *How to Lie with Statistics*, has said that the three-person and four-person families – the average – make up only 45% of the total. Smaller families add up to 35% and larger families add up to 20%. Therefore the majority of houses, etc., should not cater for average families. Also, to be average, apart from having the right number of children and just one spouse, means being the average height (important for bed length, top shelves, mirror position) with the average tastes (food storage, clothes storage, book storage) and the average requirements for water, hot baths, radiators, windows, and air. Who then is average, bearing your fat uncle in mind, or that family of six, or that man with three grandmothers, or three cars, or eighteen fishing rods? The short answer is no one, and yet authority longs to think, by the simple expedient of dividing one total into another, that all of us conform to the average picture.

Worse still, and making use of Huff again, the figures from authority can be based on false information. Curiously, and in census returns, more people are thirty-five than either thirty-four or thirty-six. (See Samuel Johnson above.) Similarly, when 100 people are asked to state the number of hours they sleep, the grand total may be a figure like 783.1. To divide this by 100, and thereby create an average of 7.831 hours a night, implies a false precision in our sleeping habits. (Over-accuracy is also false, as Johnson might have added.) Beware too of absent information. 'Cars kill 20 a day; therefore wear seat-belts.' How many of those twenty were pedestrians, cyclists or seat-belt wearers and therefore irrelevant? 'Wood-worm and Dry Rot are increasing . . . Surveys have showed that 3 out of 4 houses built before 1948 contain woodworm.' But nowhere, in this Rentokil advertisement, is evidence for the *increase*.

That prolonged London smog of late 1952 is said, time and again, to have killed 4,000. But where, in a city that has about 100,000 deaths annually (or 2,000 a week), is the evidence that the 4,000 deaths were not the total for two weeks condensed by unpleasant air into seven days? And are there not always more deaths in foggy winter than in the warmth of summer? Finally, one-third, or one-half, or even two-thirds of the world are reported as starving. To be starving is to be physically on a downhill path, with death either imminent or, at best, a few weeks ahead. Therefore are a thousand million people (or more) starving? To accept any false figure, such as that one, is to be less alert for the next fallacy, the next wrong innuendo or plain mistake. 'There is something fascinating about science. One gets such wholesale returns of conjecture out of such a trifling investment of fact,' said Mark Twain after he had calculated, bearing the river's current annual shrinkage in mind, that the Lower Mississippi must have been 1,300,000 miles long a million years ago, and would be less than a mile in length within 743 years.

In biology most facts are stated as probabilities. A fact is said to be significant when the likelihood of the event having happened by chance is 1 in 20 (or 5%). The fact becomes better, in every sense of the word, and more significant, when the likelihood of chance dictating the event is down to 1 in 100 (or 1%). Teach a dog to pick out a certain stick. If it acts correctly the first time the result is interesting, but chance could well have played its part. Everything depends upon the number of sticks coupled with the number of successful selections before it can be said that the dog is responding correctly to instruction. Even then in science there is caution: the dog 'tended to' pick up the right stick; there was a 'positive correlation' between choice and selection; the dog's actions were 'statistically significant'. Finally, and tripping gracefully over all the qualifications, the scientist will say that dogs can be taught to pick up particular sticks; but he will know that his information is based upon probabilities, not certainties, and is ready to retreat at once should further and contradictory evidence be brought to bear.

Now to the normal curve. It ought, along with an appreciation of the unlawful behaviour of chance, to be as much part of efficient citizenship – in Wells's words – as the ability to read and write. Known also as the Gaussian curve, the bell curve, the second law of Laplace, the Maxwell distribution, or even by such literary delicacies as the curve of facility of errors or the normal probability density function, the existence and shape of this mathematical function are vital. Take this eulogy from Francis Galton: 'I know of scarcely anything so apt to impress the imagination as the wonderful form of cosmic order expressed by the "Law of Frequency of Error" (yet another term) . . . It is the supreme law of Unreason. Whenever a large sample of chaotic elements are taken in hand and marshalled in the order of their magnitude, an unexpected and most beautiful form of regularity proves to have been latent all along.'

It all began on 12 November 1733, when Abraham de Moivre, an Anglo-Frenchman specializing in probability, and wanting to bring it from the gaming tables into science, published his thoughts on this subject. For a while they lay fallow until being taken up almost a century later, and most actively, by the Belgian astronomer, Adolphe Quetelet. He was a measurer *par excellence*, once totalling the weight of the combined citizenry of Brussels – about 4,000 tons – and he listed every fact he could encounter. For example, he wrote down the chest girths of 5,738 Scottish soldiers and discovered they varied between 33 and 48 inches with a peak at 40 inches. The next most frequent girths were 39 and 41 inches and this form of frequency fascinated him. When the various girths were plotted on a graph, together with the numbers of soldiers in each category, the resulting shape was something like a bell or, as some said, *la courbe de chapeau de gendarme*. This was, as Quetelet called it, thereby producing yet another name, the error curve. He was fascinated by the fact that the chest girth facts culled from all those Scotsmen made more or less the same curve as 5,738 imprecise or casual measurements of one Scottish chest.

So what is this bell of normality, this shape which outlines for us the fact that most of us are middle sized for every particular with the exceptions tapering away on both sides of that mid-line? What indeed! Gabriel Lippmann, French physicist, has said: 'Everyone firmly believes in it, because the mathematicians imagine that it is an experimental fact, and the experimenters imagine that it is a mathematical theorem.' It has even been called 'the phallic symbol of a mystique that has no foothold in the classical theory of probability'. Unfortunately, it is neither a perfect mathematical theory, although the larger the number of measurements involved the more likely these are to create a curve to fit an equation, nor is it an experimental fact. Think of the way in which idiots or geniuses, dwarfs or giants, can awkwardly prevent the smooth flow of a gently tapering curve. Perhaps such abnormalities should be discounted, but what is an abnormality? Perhaps it is merely something that disrupts the flow of a normal curve.

This curve, whatever its name and whatever its imperfections, is very important. Relentlessly it appears in the world of biometry, and virtually any measurement will express it. If the various heights (here taken from a survey of 1,323 women) are noted, and if the numbers of women falling into each category are listed, the curve immediately begins to show up, particularly if the numbers are written out in full to give a better indication of their steady growth and then their equally steady decline from the mid-point of the bell.

54″	—
55″	Two
56″	Three
57″	Eleven
58″	Fifteen
59″	Forty-one
60″	Ninety-eight
61″	One hundred and forty
62″	One hundred and eighty-one
63″	Two hundred and thirty-eight
64″	One hundred and ninety-one
65″	One hundred and fifty-two
66″	One hundred and ten
67″	Seventy-seven
68″	Thirty-four

69" Seventeen
70" Eight
71" Four
72" One
73" —

It is now important to imagine a similar curve, but one dealing with the heights of an identical number of men. As the male peak (mean) is nearer sixty-seven inches, and as men of less than fifty-eight inches are unlikely whereas men of more than seventy-two inches are likely, the whole shape will be similar but transposed to one side of the female curve. If both shapes (or graphs) are drawn on the same page, and make use of the same co-ordinates, the double image will show the male-to-female height disparity. It will be possible – and caution is necessary, for these are the murky waters of generalization – to say that men are taller than women; but by no means does this suggest that all men are taller than all women. Indeed the majority of men, if instructed to pair up with a woman of their own height, will be able to find a suitable partner. Every man of sixty-five inches or less will almost certainly be able to do so, as well as a very large number of taller men.

Neverthless it is a fact that male height, whether average, median or mode, is greater than female height. By the same account men run faster than women, even though most men have no hope of catching the fastest women. By the same token again, and these are the most disturbed waters of them all, it is possible to say that one ethnic group is more intelligent than another. Of course it cannot be affirmed which group leads if the method of testing is in dispute but, with that proviso, one large group of mankind is unlikely to be exactly equal to another large group for intelligence or any other variable. The two bell curves will not superimpose, and one of them will have its mid-line further to the left or the right of the other. Complete equality is the least probable result, and inequality is virtually axiomatic.

As with most men being able to find female partners of equal height, most members of any group will be able to find directly comparable individuals for any variable, such as intelligence, in another group. In many cases more than 99% of individuals will find their exact equal from the other camp. Yet, despite that, and wherever there is a bell curve dissimilarity, it is acceptable to state that this group is different from that group for the variable in question.

Occasionally, as with skin colour between, say, Ghanaians and Norwegians, the whitest black is blacker than the blackest white. Therefore each one of these two bell curves will begin only after the other has been concluded, but such a physical feature is not typical of the differences between groups. Generally, as with practically every other characteristic, there is overlap of the two curves, just as there is for male and female height, for speed in running, for intelligence. Racial protagonists should never be allowed to forget this point.

A final confusion, no less crucial, follows because science prefers to deal in large numbers, the larger the better, and yet we all exist in and collect our experience from small groups. Each of us knows most of all about very few people but the scientific pronouncements we have to swallow often emanate from populations. Like marry like, says science; there is a positive correlation between the respective heights of married couples. Nonsense, say the rest of us, who each know a six-footer married to a girl at least twelve inches shorter than he is. We are bound to extrapolate from our own experience, but are not always right in doing so.

Nevertheless many a scientific conclusion is reached because of someone's pre-conviction. He plans the experiment, or examines the existing figures, in the light of that pre-conviction, but he must always be ready to dismiss it if the facts, when properly examined, indicate that the conviction is unwarranted. Thirty per cent of drivers involved in accidents had alcohol in their veins, says a report. Therefore don't drink and drive, we say. But what if the same report also announces that 40% of a random sample of drivers not involved in accidents had alcohol in their veins? What then?

'The most important questions of life are, for the most part, really only problems of probability,' said the Marquis de Laplace back in 1812. If statistical thinking, as Wells wanted, were to become as important to us as mere literacy there would be less bunkum put before us in its name by politicians, advertisers and other communicators of their kind. We would also be better placed to carve our way through the most insidious jungle of all, the tangle of conflicting statements put out by the scientists themselves. Politicians and others can be arrogant in their didactic effrontery, but their methods are as thistledown when set beside the bombast and conceit of any polemic delivered in the name of science. 'Only a scientific people can survive in a scientific future,' said T. H. Huxley. Only if they know the language that scientists use; such as statistics, for example.

CHAPTER 8

Intelligence

Herrnstein – Darlington – Jensen – Eysenck – Shockley –
IQ – Genetical reaction – Intelligence and height

'Nothing in life is to be feared. It is to be understood'
                                                    MARIE CURIE

*'There is no security whatever that the individuals selected did
not have excess of character owing to environmental and not to
hereditary conditions'*
                                                    KARL PEARSON

*'If an idea presents itself to us, we must not reject it simply because
it does not agree with the logical deduction of a reigning theory'*
                                                    CLAUDE BERNARD

What has gone wrong with the issue of intelligence? In America it has
created a need for bodyguards. It has meant that lecturers have been
shouted down, long before their lectures have even begun. It has led to
physical injury, to fear, to hate-letters by the sackful. Next door a man
may be discussing the mega-deaths of nuclear conflict. Next door again
the subject may be the withdrawal of aid to developing nations, or the need
for famine, or current genocide in central Brazil; but should any academic
dare to suggest that racial differences include intellectual ability he becomes
a victim of his age. If any ordinary discussion begins to approach the subject matter of human genetics, no time is lost in the issue being diverted
to one single topic: the difference that may exist intellectually between
differing racial groups. The whole subject becomes wearily compressed
into that one inevitable issue.

Once again geneticists are not at the forefront of the fight. They were
not in the days of compulsory sterilization; nor of restrictive immigration;
and they are not today. The principal activists, Arthur Jensen, William
Shockley and Hans Eysenck, are psychologist, physicist and psychologist
respectively. To an extent geneticists are bored by the conflict. Teasing
out facts concerning the inheritance of human beings is a sufficient task

without becoming involved in the carnage of lost tempers and vanished reputations over an issue which, to them, is of marginal interest and quite insoluble. Throughout the chapter on inheritance it became plain that even single gene-pairs tie knots difficult to unravel. So too, even more confusingly, do double gene-pairs. How many gene-pairs must be involved in the inheritance of intelligence, whatever that is? And if the environment is relevant, as it is with so much of inheritance, how much more important it must be with a socially oriented phenomenon such as intellectual capability, itself involving memory, attentiveness, mental imagery, use of language, moral judgement, aesthetic appreciation, and on and on. Most geneticists, confronted with the notion of differing racial intelligence, dismiss it with two curt statements. Firstly, it is plainly impossible to get an accurate answer scientifically because the environments would have to be equally matched, and where can blacks and whites have identical environments? Secondly, even if an accurate answer could be determined, it would have no useful application.

Nevertheless, even geneticists admit the conflict does exist. First, then, to its protagonists. When they have had their say, via quotations from them and from those they have aroused, the time will then be riper for geneticists to speak out on all this other opinion. The whole science of genetics thrives on the inherited differences between individuals and between populations; therefore a digression is necessary if only to explain the considerable disinterest of geneticists in the one aspect of their discipline that, largely for reasons of history, has caught the public eye.

By way of introduction an article appeared in *The Atlantic*, an American monthly magazine of prestige and dignity. Its author was not Eysenck, Jensen or Shockley, but Richard Herrnstein, Professor of Psychology at Harvard University. He did not join their ranks; he inspected them. He gave a balanced account of the history of intelligence tests, noting how recent work fitted into the basic argument. His piece was not a polemic. It did not decry IQ tests and it did appreciate the fact that variability would continue, most assuredly.

'Greater wealth, health, freedom, fairness and educational opportunity are *not* going to give us the egalitarian society of our philosophical heritage. It will instead give us a society sharply graduated, with ever greater innate separation between the top and the bottom, and even more uniformity within families as far as inherited abilities are concerned . . . By removing arbitrary barriers between classes, society has encouraged the creation of biological barriers. When people can freely take their natural level in society, the upper classes will, virtually by definition, have greater capacity than the lower.'

Perhaps it is an affront to a society, so geared idealistically to the destruction of unfair advantages of birth, privileged class and aristocracy, to be informed that new divisions will surely take their place. Or perhaps, when so much good work is being done to equalize environments, it is merely offensive to be reminded that genetics will maintain inequality more certainly than ever before. At any rate Herrnstein was vilified. By no means had he advocated any radical or even reactionary policies; nor had he enthused over intelligence quotients, that red rag to so many. Here is how he ended his lengthy piece:

'The measurement of intelligence is one of the yardsticks by which we may assess the growing meritocracy, but other tests of human potential and performance should supplement the IQ in describing a person's talents, interests, skills and shortcomings. The biological stratification of society would surely go on whether we had tests to gauge it or not, but with them a more humane and tolerant grasp of human differences is possible. And at the moment, that seems our best hope.'

For a month after the article's appearance there was silence, more or less. Then the hate-mail began to arrive. So did the accusations of fascist and racist. His lectures were heckled and rendered less useful by scores of irrelevant questions. Students for a Democratic Society, the famous SDS, arranged a public meeting to discuss whether Herrnstein should be expelled following a rumour that one graduate had been refused permission to attend his lectures. The attacks became increasingly vehement and 107 of his Harvard colleagues placed an advertisement in the student newspaper deploring the situation: 'The open-minded search for truth cannot proceed in an atmosphere of political intimidation.'

The Atlantic itself received considerable mail. Against the article was, for example, William F. Brazziel (Professor of Higher Education, Connecticut University):

'There is little to be said about [it] except that some white editors are seizing on Arthur Jensen's article to promote the cause of white supremacy. The foreword by the editors seemed especially racist. The author of the article managed to make it through 21 pages and never say a single good thing about the mental capabilities of black people.'

In favour was, for example, S. S. Stevens (Professor of Psychophysics, Harvard):

'With the Liberal Repression still so much upon us, I was startled to see IQ and intelligence tests spread over the cover of *The Atlantic*. What has happened to the taboos? Is freedom of inquiry about to reassert itself – the freedom we once enjoyed to measure and describe the core variables that compose the human constitution? Praise be.'

Even in England, where the race issue does not simmer so near the boil, similar articles also encounter similar replies, black and white in the range they cover. Here, as further examples of the same dichotomy, are the review opinions of a book called *The Evolution of Man and Society*: *Irish Times* – difficult but honest; *Glasgow Herald* – propaganda disguised as science; *Guardian* – deliberately provocative; *Sunday Times* – tame; Alex Comfort – rashly Panglossian; Max Beloff – reasonably pessimistic; *New Statesman* – reactionary and insidious; *Morning Star* – revolutionary and often corrupt. The comments, however they might seem, were all about the same book, the most recent work of C. D. Darlington (formerly Sherardian Professor, Oxford University). He claims that in discussing issues which virtually everyone is frightened of discussing, he is having to fight on two fronts, against both the 'half-baked students and the over-baked Old Guard'. Nevertheless, as he would be the first to admit, the main field of battle is in the United States.

Arthur R. Jensen, now Professor of Educational Psychology at the University of California, is the man whose name is most intimately linked with the current storm. The subject of 'jensenism' received its first major airing in the winter of 1969. A very long article by him (it stretches for 135 pages even in book print) appeared in the *Harvard Educational Review*, and was entitled 'How Much Can We Boost IQ and Scholastic Achievement?' As learned articles go, the fact of its publication and the manner of its content were covered by ordinary newspapers and magazines with extraordinary zeal, partly because it had been distributed to many of them with a news release. The students at Berkeley were particularly active, their slogans being masterpieces of clarity: 'Fight racism', 'Fire Jensen', 'Hitler is alive and well and spreading racist propaganda at Berkeley', 'Jensen must perish', 'Kill Jensen'. Among the letters of support that Jensen received was one which, in declining to take up the issue in public, added: 'I have to admit to fears, both of what would happen to me professionally if I became identified with you, and plain gut fear of being beaten up, arson and the like. These things, if they are not here, are coming.' So what was this fascism, this incitement to violence, to murder? Jensenism includes four principal contentions. They are:

Firstly, most differences of intelligence between people are attributable to genetic rather than environmental factors, namely we are largely born not made and in the proportion of 80% to 20%.

Secondly, the races are different in their inheritance and it is 'not unwarranted' that the differential ability between Negro and white Americans in their respective IQ scores can have a genetic foundation.

Thirdly, there is evidence in America that the Negro breeding pattern is dysgenic, that those with the lowest IQs are breeding fastest. (It is also claimed that the converse is true for the white population.)

Fourthly, certain mental abilities, as well as IQ in general, are unequally distributed both among various social classes and among racial groups.

Every point has been disputed, but the 80–20 has probably been disputed most of all. Plainly intelligence is involved to some degree in inheritance, but to hazard any figure suggests some kind of ability to calculate it. Even to state the converse – that genes account for 20% – is to fall into the same pit because no firm answer, according to the detractors, is possible; the experiment just cannot be arranged that would provide the necessary results. IQ tests were established by the German psychologist Wilhelm Stern – he coined 'IQ' in 1912, the idea having originally been conceived by the Frenchman Alfred Binet in the first decade of this century – but they have been repeatedly criticized because they are claimed to measure intelligence whereas they actually measure the varying ability to accomplish IQ tests. (It has been aptly said that chimpanzees will never be proved superior to men until *they* start setting the tests.) American Negroes do score less well on IQ tests. That point is generally agreed but it is also the point where agreement comes to an end.

Jensen's opening sentence in his Harvard article was as straight and at challenging as they come: 'Compensatory education has been tried and it apparently has failed.' He feels the reason why the backward have not been brought sufficiently forward by efforts to make good the gap is that IQs have largely been inherited. The die, in short, has already been cast from the moment of conception. The environmentalist counter-argument is that those with low IQs come from poor homes, poor backgrounds. They donate these factors to their descendants, thus maintaining the low level of IQ, rather than any genes causing low IQ. The situation is one of stalemate, with each side fully entrenched. Even if Jensen's point about compensatory failure is accepted the environmentalists merely reply that compensation should have begun earlier, before schooling started and when the effects of a poor home are at their most deleterious. It is an impasse, and one from which most people, including geneticists, have largely withdrawn.

Not so psychologist Eysenck; not so physicist Shockley. The principal battering so far given to Hans Jurgen Eysenck ('not one of my paperbacks has sold less than a million copies') was at the London School of Economics in May 1973. Pulled to the floor, punched, kicked, and spat on, he was then dragged, with glasses broken and nose bleeding, to safety. His crime, in effect, had been to state, as with Jensen and Shockley and others, that intelligence is 80% controlled by the genes. When questioned at a lecture he was prepared to admit to 70% or 90% but not to the creed that such figures are both meaningless and harmful. There cannot, he says, be the socialist utopia, with everybody receiving the same opportunity; therefore today's socialism is in a dilemma, having either to give up this ideal or else to renounce the facts.

A review by Eysenck of the book *Race, Culture and Intelligence* (edited by Ken Richardson and David Spears) helps to explain his views:

'I was particularly pleased to note that I could detect almost no differences of substance between my position and that of W. F. Bodmer (Professor of Genetics, Oxford University), who contributed the crucial chapter on race and IQ. In particular, we seem to arrive at the same conclusion about the social consequence of the debate, and I shall close by quoting his last few sentences. "Our accepted ethic holds that each individual should be given equal and maximum opportunity, according to his or her needs, to develop to his or her fullest potential. Surely innate differences in ability and other individual variations should be taken into account by our educational system. These differences must, however, be judged on the basis of the individual and not on the basis of race. To maintain otherwise indicates an inability to distinguish differences among individuals from differences among populations." This presents exactly what I tried to say in my own book, if less elegantly and less persuasively than Bodmer.'

Without in any way forgetting the argument, but changing the arguer, there is also Shockley, a man notably more strident than Eysenck. More than the others, and despite an occasional need for bodyguards, he welcomes the controversy. 'If you can't stand the heat,' said Harry Truman, 'keep out of the kitchen,' and there is no sign of Shockley leaving any room. For example, when confronted by 1,800 students at Stanford, many of whom were black scholars, he suggested in his opening remarks that they should take part in a research experiment to test his 'tentative prediction that for each 1 % of Caucasian ancestry the average IQ of American black populations goes up by approximately one IQ point'. The meeting lasted for four hours.

Leeds University elected in December 1972 to award an honorary doctorate to Shockley for his work in helping to invent the transistor, work that had already been honoured (in 1956) by a Nobel prize. Two months later the university changed its mind and cancelled the invitation; the reason was the professor's recent pronouncements on heredity. Shockley has said that he became interested in the subject because he had found an 'intellectual vacuum'. He has compared it with 'the irresponsibility that existed in pre-war Germany when intellectuals were unwilling to face up to the horrors that might be happening'. After investigating the intelligence situation he was 'led inescapably to the opinion that the major causes of the US Negro's intellectual and social deficits are hereditary and racially genetic in origin, and thus not remediable to a major degree by any practical improvements in environment'.

This is, in essence, jensenism. What is shockleyism is his extension of that hypothesis, an idea that has also been called the American final solution. Briefly, it is to offer financial incentives to all those with little intelligence so that they produce fewer children. Those who volunteered for sterilization would receive a cash bonus related to a scientific estimate of their disadvantageous hereditary factors, 'such as diabetes, epilepsy, heroin addiction and arthritis'. On intelligence Shockley has considered as 'a thinking exercise' that sterilization should be accompanied by a $1,000 award for each IQ point below 100. As it has been alleged that Negroes are fifteen points below the white average of 100 this would indicate, if the thinking became a practical exercise, that the average Negro would receive $15,000 as his country's grateful reward for ducking out of its breeding programme.

My motives, says Shockley, are humanitarian.

'The view that the US Negro is inherently less intelligent than the US white came from my concern for humanity ... The failure of the intellectual community to deal with these matters is one of the cruellest irresponsibilities to a minority group that has ever occurred. If, in the US, our nobly intended welfare programmes are indeed encouraging the least effective elements of the blacks to have the most children, then a destiny of genetic enslavement for the next generation of blacks may well ensue. It is my considered opinion and evaluation that, at the present time, I am less likely to do damage by exacerbating a situation, and am currently the intellectual in America most likely to reduce Negro agony in the next generation.'

So much, briefly, for Jensen, Eysenck, Shockley and co., plus a sufficient crack of their various whips. Now to the other side of the fence, or rather to opposing opinions from various sides.

In essence, sociologists and others of their kind are so fascinated by the environment and what it can do that they swell up with resentment at genetics, at being told what the genes are doing, have done, and will continue to do. Conversely, the environment is the curse of the genetical classes; it is so ready to interfere with any experiment that scrupulous care must be exercised to keep it constant. If it varies, the genetic consequences of an experiment are immediately rendered less valuable; environment has had its say. Hence the kind of abject, nonplussed confusion evoked in geneticists by much of the current talk about racial intelligence.

Here is one determined attempt by a geneticist at a rational rebuttal of that notorious 80–20 ratio.

'Every student of genetics knows that estimates of heritability are applicable exclusively to the population and environmental situation for which they were calculated. The famous 80% value was derived from studies on a white population, mainly of a middle-class origin. Even if we accept it as correct, it still applies only to that population. No similar investigations have been carried out on Blacks or Red Indians. Those who would like to promote Professor Shockley's programme of bribing people with low IQ into having themselves sterilized must restrict these measures to American Whites ... The sooner we forget about the 80% – or any other – value of heritability of IQ, the better.'

A basic point of disagreement concerns the IQ tests themselves. Even Eysenck (in his *Know Your Own IQ*) has written: 'Intelligence tests are not based on any very sound scientific principles, and there is not a great

deal of agreement among experts regarding the nature of intelligence . . .'
There is even less agreement regarding the nature of the results when IQ
tests are made to reveal racial differences because, if races exist, it is virtually axiomatic that different cultures also exist, whether at a distance or
within the same geographical area. The genes are different; so too the
cultures and the environment.

It may be refreshing, and even relevant, to recall that tale of a researcher
who, in some dark corner of a foreign land, was wishing to learn if its
inhabitants knew about two-dimensional perspective. He had a drawing
of a man with his spear arm raised and two animals in view. The first, an
elephant, was close at hand, an easy target. The second, an antelope, at the
far end of a tapering highway, was clearly and impossibly beyond the
hunter's reach. The interviewees would reply, time and time again, that
the antelope was the animal most likely to die; it was plainly the point
of the hunter's aim. The interviewer, who knew that an elephant was a
more desirable quarry, and who had slanted his drawing with that fact in
mind, was overjoyed. By his questioning he had stumbled across a gross
error in visual perception, and he prepared to depart. Just in time he overheard: 'What a fool would be the man who, when on his own, would
throw one spear at an elephant.' IQ tests are also imperfect, particularly in
different environments. The conclusions they produce are important, but
should always be approached with caution.

Of course it is true that inheritance is involved in intelligence. As one
man wrote to *The Atlantic* following Herrnstein's piece: 'The assumption
that our minds should be exempt from the genetic laws that govern the
potentials of our bodies has always seemed to me absurd.' On this point
Jensen went a little further: 'Genetic differences are manifested in virtually every anatomical, physiological, and biochemical comparison one can
make between representative samples of identifiable racial groups'; therefore 'there is no reason to suppose that the brain should be exempt from
this generalization'.

The genetical reply to this point, as in the textbook *The Genetics of
Human Populations* (by Cavalli-Sforza and Bodmer), is terse: 'There is
no reason why genes affecting IQ which differ in frequency in the gene
pools of blacks and whites should be such that, on the average, whites
should have significantly higher frequencies of genes increasing IQ than
would blacks.' A gene pool is, as its name implies, the total collection of
genes possessed by an entire population, whether white, black or living
on Tristan da Cunha. The pools may be different, and the frequency of

certain genes within those pools may be different, but that is no reason to presume the genes will add up to a higher total intelligence in one or other population. In fact, the more polygenic a feature, and undoubtedly scores and scores of genes must have their say in creating the basis of intelligence, the more their effects will smooth out any specific differences between two groups/populations/races.

In the absence of all actual information on this score it must consequently be assumed that natural selection for intelligence has operated evenly on the major groupings of mankind. The history of man gives no evidence that intelligence (or rather IQ, as no one has yet measured intelligence) has been more favoured for survival in any one large community than in any other. Therefore, if it is expected that no genetic difference in IQ between the races exists, the focus must swing back to the environment.

Most certainly environments do differ. Within any community of rich and poor, privileged and under-privileged, stimulated and aggravated, there is no uniformity. Even the uterine environment is not the same (or why do the poorer classes experience more congenital malformations of the central nervous system among their offspring?). The family environment is not the same (or why does the child of a large poor family fare less well in IQ tests if he has many siblings above him?). The health situation is not the same, nor are the educational opportunities. Above all, and as the most undeniable fact of racial difference, no one can pretend for an instant that, in black and white America, these two worlds are the same.

Consequently, as the geneticist has to look first to the environment before he can begin to conduct a genetic experiment, there is only one possible method for studying racial IQ. Black children would have to be adopted into white homes, and vice versa, preferably on the day of birth. To reduce the normal parental effects, the mixed adoptions would have to be partnered by a similar number of homogeneous adoptions as controls, of white infants into white homes and black into black.

The striving geneticist, aiming so hard to extract some valid information, would quickly find that what is possible in a laboratory is impossible outside it. The human trial would immediately encounter the very conflicts it was attempting to disentangle, for a white child in a black neighbourhood or a black one in a white world would not merely have swopped environments. They would both, at the present time, have met prejudice. They would be experiencing a twilit existence, neither one nor the other. 'We therefore suggest,' L. L. Cavalli-Sforza has written, 'that the question whether there is a genetic basis for the race IQ difference will be almost impossible to answer satisfactorily before the environmental differences between blacks and whites in the United States have been satisfactorily reduced.'

The dilemma is real. Should the geneticists attempt work for which they feel there is no good case on either theoretical or practical grounds? Should they even enter the fight? First, Cavalli-Sforza: 'There can be no doubt that in the present racial climate of the United States, studies on racial differences in IQ, however well intentioned, could easily be misinterpreted as a form of racism and lead to unnecessary accentuation of racial tensions.' Second, and conversely, Professor Dobzhansky: 'Faced with a revival of "scientific racism" one is tempted to treat the matter with the silent scorn it so richly deserves. The temptation must be resisted, however, lest some people should be misled by a spurious vindication of race bigotry.'

The dilemma will only vanish when the current racial tensions have gone for good. In the meantime every fact, even half-facts and hypotheses, hedged about with qualifications in the normal scientific manner, will be exploited far beyond their worth. That 80–20 notion, postulated by some and disbelieved by others, has achieved a fame out of all proportion to its validity. On it men have piled opinions. From it men have extrapolated wildly. Bad science and bad politics have an affinity for each other that is always alarming, and frequently explosive.

By way of adding some leaven to the conclusion of a fairly weighty piece here are some positive statements linking intelligence with stature. The trouble with race and intelligence, as is now abundantly clear, is that positive scientific statements are in short supply; hence the continuing argument. In the hope, therefore, not only of producing a few facts but of providing some substance for further argument along a less downtrodden path, these additional remarks may help to provoke new thoughts and possibly fresh antagonism concerning inherited ability.

'We can confidently assert,' Professor J. M. Tanner has written, 'that there is currently a small but significant tendency for taller adults in the population to score higher in some intelligence tests than short adults of the same sex . . .'

Although linked with correct qualification the meaning is clear: to be taller is, on average, to have a higher IQ. Bring that fact up the next moment anyone dares to disgorge the aged talk of racial inequality. Here are some more from the same author who took pains to collect the relevant material.

'It seems likely that tall people rise (in the social scale) because they are marginally better at certain mental tasks.' 'With schoolchildren ... those that are larger score more highly than smaller children of the same age ... The chance of a large child passing an exam is substantially greater than that of a small child.' 'This correlation diminishes when maturity is reached but does not disappear entirely.'

Women are not exempt from these general dicta and they too can be swiftly brought into the discussion. For instance, here is Tanner again:

'On average taller women tend to rise in the social scale, both in getting jobs and in marriage; and shorter women, on average, tend to sink. This is probably due to height being related to ability.'

Perhaps that is sufficient. Should the argument reach violence, or should the back-up material for these statements be demanded, may I recommend that Tanner's excellent paper is read in its entirety. It will be found in *The Eugenics Review*, volume 58, Number 3, pages 122–35. Reading it may provide a double blessing; it may also help to show why biologists and geneticists are always interested in hypotheses that can be tested for their worth and heavily disinterested in that threadbare topic of differential intellectual ability between the races.

CHAPTER 9

People numbers

Population growth – Compound increase – 3% or 1% –
T. R. Malthus – Fears for fertility – Fears for increase –
A right to breed – Financial inducements – China and
Singapore – America's phenomenon – Future outlook

*'The practices [of the limitation of families] are responsible for a
deadening of moral sensibility, and for a degradation of character
among those who resort to them'*
ROYAL COMMISSION ON THE DECLINE OF THE BIRTH
RATE SET UP IN NEW SOUTH WALES IN 1904

*'They, the population "doomwatchers", have been in existence
for more than a century – in fact, since the time of Malthus.
I do not think we should pay them any more attention now than
was paid to them then'*
EDWARD HEATH (1972)

Before there is any kind of quality control of human beings there must of
necessity be quantity control. At present both are almost equally uncon-
trolled, but a kind of dawn is beginning to break in the attitude of our
species towards its numbers. By no means has this enlightenment become
universal but things are changing, there is a different feeling in the air and
quantity control no longer evokes the antagonism that once prevailed.
In time quality may be subjected to the same pressures and it too may be
toppled from its current status of laissez-faire. No one knows when or how
this will happen, but an examination of the recent events concerning quantity gives an indication of, among other things, the possible swiftness
of all future change. Examine our attitudes to quantity change, therefore,
but keep the idea of quality change somewhere near the forefront of your
mind.

Population is a splendid subject. It enables all of us to express concern,
to extrapolate, to point the finger at other countries, to do arithmetic with

147

startling results, to wonder where it will end when we are safely dead. Population is other people, like traffic, like the public, like a crowd. Besides, or so it readily appears, every statement on the subject can be confronted by some opposing and equally startling contradiction. Contraception must be the key to family planning, we say, and yet the British birth rate started a long decline in 1870, some forty years before contraception, in the form of rubber sheaths, came cheaply on the market. We have no population policy, we say, and yet have tax rebates, family allowances, maternity benefits and all manner of economic links associated with the possession of children. Appalled by today's apparent vindication of his prophecy we nod our heads at Thomas Malthus, and then hear Herman Kahn insisting that a doubling of the population will mean a quintupling of the world's economy. Even that word explosion becomes suspect when we later realize it means an annual growth of 2%, a modest increment much like the annual expansion of any tree.

Perhaps a start could be made by handing out, instead of free advice or literature, those small calculators that instantly flicker their electronic answers to any problem punched into them. First, and as a useful sum, there is the potentiality of the breeding system, considerable in mankind, but far faster in mice. Assume two new-born mice as a beginning. Know that they are capable of producing eight offspring within ten weeks. Realize, as two parents are required for each new family, that this means fifty mice in twenty weeks. And then be amazed, or complacent, or horrified, that this could mean twenty million mice in two years.

Humans are undoubtedly slower, but not so laggardly as might be supposed. Take two human emigrants of the early seventeenth century travelling to a new and empty continent. Assume this couple to produce a family of eight, neither large nor small in those earlier times of emptier lands. Assume also, for the sake of simplicity, that all eight of their children both marry (each other initially to make the sum easier) and then have a further eight children per marriage. The emigrant pair will therefore be able to smile upon thirty-two grandchildren before they take their leave. However, assuming their tradition of eight children per couple is carried on faithfully, and knowing there has been time for thirteen generations between that early emigration and today, the original pair could alone have been responsible for current progeny totalling 134,217,728.

Mice would have done better, but the result is impressive, nonetheless. Remember, by way of a practical illustration, the story of the French in Canada, the so-called revenge of the cradle. With their defeat by the

British at Quebec in 1759, and the immediate curtailment of further emigration from France, the 7,000 French within the colony licked their wounds and began to breed. There are now 3½ million in Canada who speak only French, and a further 2 million who are bilingual; not bad for only two centuries and 8½ generations of cradle revenge.

The pocket calculator, responding so tacitly to one's every inquisitive whim, is equally forthcoming about the colossal effects of relatively small differences of behaviour between groups. Assume group A, underdeveloped and traditional, marries when young and produces four children per family. This seemingly unexplosive and reasonable behaviour will lead, from a start of one new-born couple, to a population of six after 25 years, of twenty-two after 50 years, of eighty-four after 75 years, and of 1,344 after 100 years. (Assumptions here are of one generation every twenty years, and a life expectancy of fifty.) Group B, developed, industrial, marrying five years later, living until seventy, and producing three children per family, leads to very different results within a similar breeding span of 100 years. After 25 years the original new-born couple have become five; after 50 years their numbers are fourteen, after 75 years they total thirty-nine, and after 100 years they add up to only 117 people.

Therefore group A, by having one more per family and marrying five years earlier, produces 1,344 in a century as against 117 for group B. This is a phenomenal differential bearing in mind the fairly small difference in behaviour. The actual time of death, providing it happens after each individual's reproductive span, is not important. If all those undeveloped citizens of group A were to die on average at 35 (very low even for today's poor countries) the original couple would still create 1,280 people by the end of a century. On the other hand, if all the citizens of the wealthy group B discovered an elixir to make them immortal this fact would also make little difference: the original couple of group A and their descendants would, assuming this immortality, all be alive at the century's end but would only total 122.

In short, the fact of early marriages is as powerful – for exploding a population – as the relentless wish to have one more child. A practical illustration of this point, giving an understanding of the relative stability of Britain's population in the seventeenth and eighteenth centuries, is contained in the parish registers. The average age for marriages then, complicated by the rules for starting married life with certain cash or property assets, was much later than now. Only three generations per century then occurred as against five in many of today's communities. With the first there is stability; with the second an explosion.

The prevailing demographic custom is to tabulate average life expectancy at birth without any reference to the breeding span. Once again it is a mean figure that is at fault because Indians of India, for example, do not tend to die near their current average of forty-eight. They die, largely, either in infancy or at old age. A child dying aged 4 and a man dying at 92 give, entirely misleadingly, that mean of forty-eight. Demographic statements would make much more sense if they indicated the proportion of live births expected to reach puberty and therefore liable to add to the population.

The helpful calculator will also happily demonstrate compound interest to those of us either unappreciative of its inexorable momentum or too idle to do the arithmetic. An increasing population proceeds in a compound fashion, just as savings are supposed to grow, and the manner of its increase should be part of our understanding. Three per cent growth per year means, of course, that 100 becomes 103 in twelve months. It also means that 103 becomes 106.09 one year later, with 0.09 being the trifling advance on 106 that has all the virulence of the first snowflake to start an avalanche. One year later, following the same 3% growth, the figure has been compounded to 109.27, still trifling but the snowflake is on its way. A steady punching of the calculator, a steady increase of 3%, leads to a doubling of that initial 100 after only twenty-four additions (or yearly increments) of that percentage. By then the avalanche is gathering speed. The second hundred is reached after fourteen more additions of 3%, the third hundred after nine more, the fourth after eight more, the fifth after six more.

So with people. At 3% a year 100 people become 200 after twenty-four years, 300 after a total of thirty-eight years, 400 after forty-seven years, 500 after fifty-five years, 600 after sixty-one years and 700 after sixty-six years. In other words, a mere 3% growth rate – a figure exceeded throughout Latin America – transforms 100 million people into 700 million people within less than a lifetime. If only the growth rate would drop to 1%, making that first year's increase push the population up to 101 instead of 103, the number of people would not even double in sixty-six years, let alone multiply sevenfold. The calculator is quick to put us right, however poor at arithmetic we may be, and however difficult it is to understand where – from diseased, malnourished and under-privileged lands – all those extra mouths are coming from.

The world in general is actually expanding its people at 2%. Europe, the USSR and North America are below that percentage while Asia, Africa and South America are above it. With Europe well aware that it is the most crowded region of them all, with talk from Russia of one-child families, and with Americans discussing zero population growth, it would seem that the slowest expanders are the most concerned about the problem. Therefore, genetically speaking, the kinds of people that have in the past been breeding most energetically are no longer doing so. A new pattern of inheritance is now in train, most speedily, and new kinds of people are making the pace.

Until now, as with the problem of who marries whom, there have been various laws, edicts, social customs and mere happenings controlling the numbers of offspring, but little that could be called a policy. Bearing in mind those frightening extrapolations of undisciplined breeding, and of the overcrowding, famine, pestilence or war likely to result, it would seem that a policy for controlling quantity is frighteningly crucial. At long last the revulsion felt for this form of control is on the wane, and the swing has been quite sudden. Remember that any interviewer (say on radio, or in the village hall) could – just a few years ago – feel certain of a round of applause if his interviewee proved to be either old (over seventy) or abundantly fertile (with six or more children). Today the old will still be applauded, and perhaps even that custom will die, but there is no similar guarantee for the prolific. Something has changed in us; the possibility of a population policy is nearer by far than ever before.

A few centuries ago, apart from a belief that children were good, and the more the merrier, there was a general sentiment that the state should encourage reproduction but not concern itself at all with the result of that encouragement. For example, there was Sir John Harington writing 300 years ago in England that the fathers of ten children should pay no taxes while all bachelors over twenty-five should pay double the existing rate. In the following century a bill for taking annual censuses in Britain was passed by the House of Commons (in 1754) but rejected by the House of Lords. There was strong religious feeling against 'the numbering of the people', and their lordships believed the lower classes would become restless if an act contrary to divine will was imposed upon them. It was for this reason that Thomas Robert Malthus, the timid ex-Cambridge don with a hare-lip, thought it better to publish his famous thoughts anonymously. He had recently become a country curate in Surrey, and assumed his subject matter would be a cause for scandal. Instead his words opened up a new era in population concern.

The Essay on the Principle of Population as it affects the Future Improvement of Society appeared in 1798. He wrote:

'I think I may fairly make two postulata. First, that food is necessary to the existence of man. Secondly, that the passion between the sexes is necessary, and will remain nearly in its present state ... Assuming then, my postulata as granted, I say, that the power of population is indefinitely greater than the power in the earth to produce subsistence for men. Population, when unchecked, increases in a geometrical ratio. Subsistence increases only in an arithmetical ratio.'

It is for the general truth of this dictum that he is remembered, and not for his further statements (published under his name in 1803 and then again in 1830) concerning the dangers of poor relief, the inevitability of poverty and private property, the harm of deliberate birth-control, and the possibility of revolution (from all those unrelieved, un-birth-controlled poor) but at least he had spoken out about the implications of population increase. More extraordinarily, he had done so at a time when population numbers were considered all to the good. Britain's great rival of the time was France, with her 28 million people as against the British 11 million, but Britain was catching up and was able to profit in an expanding world by her expanding numbers. Nevertheless, however timely for Britain that her birth rate had suddenly been able to take significant strides in advance of her death rate, it was the beginning of a new age and Malthus had been able to peer ahead for some of its consequences.

The first major step taken by Britain to assess her changing population did not occur until a century after Malthus had been buried. This was prompted not by the excessive expansion that had concerned him, although the population was seven times greater than in his day, but by its possible contraction. For reasons no one could fathom, and still cannot, the birth rate altered its upward curve in the 1870s and began a downward trend that lasted for over sixty years. Was it the depression of those days? Was it the initiation of free education, suddenly giving parents hope for the future of their offspring? Was it the famous Bradlaugh-Besant trial of 1877 that gave such widespread publicity to the idea that control of conception was possible? Was it the growing emancipation of women? Was it linked in any way to tremendous emigration – 50 million left

Europe between 1846 and 1932, with 18 million from Britain alone? Or
was it just that the birth rate could not ascend for ever and would, by the
cyclical nature of events, be bound to change sometime, with the 1870s
being as good a time as any?

Not until 1933 did the pendulum begin to swing back. At that time,
to confound those proposed conjectures in the previous paragraph, there
was a far greater depression than ever before, free education had been
both improved and lengthened, conception control was better understood,
women even had the vote, and only emigration was behaving differently,
with more nationals returning than leaving.

The general concern about population decline continued even after the
birth rate had started to rise once more. With the arrival of World War II
in 1939, with its loss of young life plus the separation of young couples,
concern rose appreciably. A Royal Commission on Population was set up,
the first in British history. Its conclusions were not published until 1950
and, to a world now immured with talk of a population deluge, some of
its conclusions (from such a recent year) make arresting reading.

'If in future British fertility rates remain below replacement level, as
they have been for many years past, the population of Great Britain,
unless sustained by immigration, will presently cease to increase and
begin to decline . . .

'The economic consequences of the prospective future trend of popu-
tion also have a bearing on the desirability of intervention by the State
to promote a recovery in the birth-rate . . .

'There can be no question as to the necessity of restoring our fertility
rates sooner or later to replacement level, since the alternative is national
extinction . . .

'A smaller population would be advantageous on balance of payment
grounds, but it is difficult to forecast whether this advantage would be
slight or important. There would also be a gain on the score of amen-
ity . . .

'We conclude accordingly that in Great Britain at the present day the
case for reasonable and well-considered measures to mitigate the burden
of parenthood is fully made out on economic and social grounds.'

National extinction! State intervention to boost the births! Much of it
sounds as if from another century and another world, let alone our own
community and in the lifetime of most of us. Nothing was done to follow

up the recommendations, partly because such commissions are often an end in themselves, taking the steam out of the much-felt need to examine a situation, but largely because – in 1950 – the country was in the grip of a baby boom far greater than that experienced after World War I. In the 1960s, when the global picture was better understood, when the wartime weapon of DDT was being used against the mosquito, and when the under-developed countries suddenly leapt into population prominence, the world knew it had a people problem on its hands. Someone called it an explosion, and countless modern Malthusians have also made much noise about the extraordinary change in our long history upon the planet. 'If we don't do something dramatic about population and environment, and do it immediately,' said Paul Ehrlich, biologist, propagandist, father of one and now sterilized against a further such occurrence, 'there is just no hope that civilization will persist.' What is the optimum population, we ask? 'We have already exceeded it, gentlemen,' said John H. Knowles, director of Massachusetts General Hospital, 'we have already exceeded it.'

Once again there are contradictions making it harder for us, all 3,700 million of us, to be united in our opinion that there is a population problem. In the 1960s half of America's 3,000 counties *lost* population. In the same decade many South American countries suddenly started to realize their national potentials, and politicians spoke warmly of their future power and authority in the world when their people were ten times more numerous. Africa is still empty – by any European yardstick. And are not the problems of big cities, with overcrowding and clogged highways, more the result of bad management than gross numbers? Were the cities not crowded, foul and clogged a century ago? Is it not a problem of excessive exploitation than of excessive numbers when America devours 40% of the resources for 6% of the world's people, when each fresh birth in Europe is twenty times more demanding upon the planet than each new Indian?

There is also resistance to the idea of national, let alone international, breeding control. Is it not some kind of birthright, we say, much like air to breathe and water to drink, with every partnership of two people producing children as they and their god see fit? 'We like sex, and we like children,' is a cry right round the globe. 'The force of habit is a terrible force,' said Lenin, and no one would disagree. 'Most people who anguish over the population problem,' as Garrett Hardin said in *Science*, 'are trying to find a way to avoid the evils of over-population without relinquishing any of the privileges they now enjoy.'

Certainly there is no rush to relinquish anything. In 1967 a statement was made jointly by thirty nations: 'The Universal Declaration of Human Rights describes the family as the natural and fundamental unit of society. It follows that any choice and decision with regard to the size of the family must irrevocably rest with the family itself, and cannot be made by anyone else.' It is painful, said Garrett Hardin, 'to have to deny categorically the validity of this right'. 'We want every country to have a population policy,' wrote Julian Huxley (in 1964), 'just as it has an economic policy or a foreign policy. We want all the international agencies of the United Nations to have a population policy . . . The control of population is a prerequisite for any radical improvement in the human lot.'

It is so easy to make population pronouncements, and certainly American presidents, for example, have been repeatedly happy to make them. 'The population problem has become one of the most critical world problems . . .' said President Eisenhower. 'The magnitude of the problem is staggering . . .' said President Kennedy. 'I will seek new ways . . . to help deal with the explosion . . .' said President Johnson. 'One of the most serious challenges to human destiny . . . will be the growth of the population,' said President Nixon. And yet no policy has emerged from any American administration – thus far. Neither of the last two British governments has advocated anything on the subject, believing it too hot for safe political handling.

Conversely, the actual promotion of births has been at times politically acceptable and various countries have introduced schemes. France forbade even birth-control literature to be circulated until very recently and has given welfare bonuses for large families. Czechoslovakia, since the Russian invasion, has offered housing concessions – low rents or better loans – to young couples if they have more children, and the government is congratulating itself on the boosted birth rate. Pre-war Germany undoubtedly aided births by hindering abortions, promoting motherhood and giving tax incentives.

Future generations may wonder why today's governments are quite so cautious about population control, as they already exert considerable influence on our reproduction rate. After all, and without a whisper of a population policy, the financial incentives existing in many countries cannot be irrelevant. Take the British system. There is income tax relief – £240 annually for each child under eleven, rising to £275 for 11–15, and to £305 if sixteen or over and still being educated. There is also a 'family allowance', not (strangely) for the first child, but of £1.50 a week for subsequent children, although this allowance is subject to tax for the wealthy. There is a maternity benefit for each confinement of £176 (provided the mother has paid *her* insurance contributions), and a maternity grant of £25 for each child. As further aids to parenthood there is free ante-natal and post-natal treatment for the mother, plus free dental treatment when she is either pregnant or the child is less than one. For the child there is free medical treatment until he or she leaves school, free dental treatment until the age of twenty-one, free milk, cod liver oil and vitamins for the needy, and free education. Finally, both the allocation of council (state) houses and of (state) income supplements relate closely to the numbers of children involved in each family.

By no means are these aids, allowances and grants of milk and oil sufficient to pay for any child's upbringing, as all parents will agree who have to purchase so many 'necessities' when their children mewl, puke, crawl, walk and run through life (and much parental cash); but any incentive, any relaxation of the financial stringency, must have an effect. Few parents can tabulate how much each child is costing them, but they know that the overall burden is, however confusingly via tax and allowances, and those taxed allowances, less than it might have been. Moreover, we are not logical over money. We do rush in if something is free, or half-price or even reduced. We are bargain seekers. We will travel all over town to find good value, and fail to count the cost of all that travel correctly. No family will accept the idea that little Henry or young Jane was the result of an improved family allowance but, for the nation as a whole, governmental spending on such rebates must bring its own numerical reward.

China, so far, is the only large country to have a positive population aim. Other countries have wishes, with India wishing most of all, but only China has announced a fixed intention, a reduction in its population growth to an increase of $1\frac{1}{2}$% a year by 1980. Chou En-lai, the prime minister, said in 1971 that growth was still 2%, meaning a yearly increment of sixteen million on the 800 million already achieved. If $1\frac{1}{2}$% is realized by 1980, and if the 2% is steadily whittled away before then, China's total will still be a formidable 930 million by the start of the next decade. Every fifth person on the planet will then be Chinese and the republic's population will still be expanding at $13\frac{1}{2}$ million a year. Therefore China is likely to be the first nation, beating even India, to reach a thousand million people, or the population of the entire planet back in 1850. The Peking government is making tremendous efforts to control this boom, by handing out free contraceptives to every woman when she collects her monthly sanitary towelling, thus getting directly at the breeding community, but the momentum is considerable. That addition of $1\frac{1}{2}\%$ a year would have to drop well below even the European increment of 0.7% if China is not to achieve her tremendous gathering of 10^9 inhabitants.

Singapore has been even more forthright, and is the first community of any size to be entirely frank about using hard cash to soften the population problem. The country began its family planning programme in 1966, mainly with slogans. The birth rate dropped from 28.6 per thousand in that year to 22.1 in 1970. Afterwards, and remorselessly, despite more slogans from the government, it began to creep up again. By then the republic's 225 square miles, equivalent only to a square with 15-mile sides, contained 2,100,000 people, or one person in every 330 square yards.

Something more than coercion was needed, and cash was chosen. From 1 August 1973 it was decreed that maternity leave would not be paid after the third child, that the tax-free child allowances would be abolished and hospital charges greatly increased for the fourth and all subsequent children. Moreover, the birth of a fourth child would mean the guilty family going back to the bottom of the government housing waiting list. The new laws were announced exactly nine months before they were due to take effect, and therefore with inadequate biological time for parents to rush through another child with the old benefits. Nothing in the new laws is retrospective – existing large families are not to be punished – but the experiment will be eagerly watched by the world, and not least by the Government of Singapore.

Without doubt, and genetically speaking, the future looks increasingly Asiatic. South America is breeding faster, doubling every twenty-five years but only from today's modest figure of 276 million. Africa is expected to double in twenty-eight years, but only from 345 million. Asia, relatively laggardly and with a doubling time of thirty-three years, has the formidable head-start of a current population near 2,000 million. Already more than half of this planet's people live in Asia, and there is every indication that this proportion will increase as the years rush by.

The greatest amount of talk about a stable society, and zero population growth, has come from the United States. This is odd, bearing in mind the considerable celebration when America's population clock in the US Department of Commerce at Washington ticked for the first time past the 200 million mark (in November 1967) and the fact that Americans are not all that crowded. (There are 58 of them per square mile – about one-sixth of the density of Switzerland, or one-fifteenth of that of England.) The United States has for so long seen itself as the New World, the wide open space, the free for all. Also the pantheistic concept of an economy geared to growth has had a greater following in that country than in any other nation save, of course, for post-war Japan where GNP are the most famous initials of them all. Perhaps the American method of living makes the place seem more crowded than any back-street in Bombay because Americans throw away a million cars a year, fifty-eight million tons of paper, 36,000 million bottles, and jam 70 % of themselves into 2 % of the available surface. Many Bombay Indians would not be able to fill one garbage can throughout their lives. The New York, Baltimore and Philadelphia area is a crowded, clogged conglomerate, whereas Bombay and Calcutta just have a lot of people. Moreover, America had only fifty years to acccommodate her second gain of a hundred million Americans and has begun to feel fearful, however great the technical resources, of the social consequences of a third such gain in even less time. As Paul Ehrlich put it, never missing the opportunity for a powerful passage: 'America's pride in her growing population may be compared to a cancer patient's pride in his expanding tumour.'

The trouble with any reasonable attempt to achieve a stable population is that it takes a formidable time to achieve. If Americans were to maintain current immigration (400,000 a year) but limit every resident family to two children, the population would still grow to 266 million by the end of the century and 321 million by 2070. Yet more forbidding, a three-child average (with the same immigration rate) would create nearly 1,000 million Americans a hundred years from now. To achieve zero growth immediately would be to limit families to replacement level, to compensate only for deaths (after allowing for immigrant arrivals). The average family would then have less than one child. If immigration were stopped entirely, and if Americans were limited to two-child families, the population would continue to grow until 2037 and would then total about 270 million.

In an excellent report submitted in the late 1960s by the Commission on Population Growth and the American Future (chairman John D. Rockefeller III) it is strongly affirmed that 'the time has come to ask what level of population growth is good for the United States'. It first enumerates the problems – natural family inclinations, immigration requirements, shrinking cities with growing suburbs, increasing awareness of the hazards to come with rising levels of consumption per head – and then adds: 'The choice among ways to redirect growth does not eliminate the necessity of making the choice about when the Nation could best accommodate 300 million people or whether it should accommodate 400 million.' Its simplest finding is that the prevention of all unwanted births within the United States would virtually lead to zero growth overnight. Can such a grave problem really have such a beautiful solution?

However, and most remarkably, the problem suddenly appears to be in danger of vanishing. The 1970 US census revealed a unique event. The record *increase* among young adults in the most fertile age group (known about ever since the baby boom after World War II) has coincided with a record *decline* among children under five years old. Despite a fairly high immigration, basically of young people, and despite all fears that the biggest baby bulge of all time would mimic its own boost to the population, the result has been entirely contradictory: the pre-school-age population of the US declined by 15.5 % between 1960 and 1970, a phenomenon unmatched in the history of America.

Normally this age group has grown somewhat in keeping (wars and depressions permitting) with the general population growth, but the highly fertile 20–24 age category increased during the 1960s far more rapidly than at any time in the past and yet has brought about a bigger relative drop than at any other time. American girls are just not having babies at the expected rate. In fact, at the last census, only 78 women out of every 1,000 between fifteen and forty-four – the breeding era – had babies, as against 90 in wartime 1942, or 107 when the Korean war began, or 123 in 1957, the most fertile US year in recent times. As a consequence of this drop in the production of babies the average family size for newly-married American couples is only 2.04 children, as against 2.39 two years ago, and as against 2.11, the figure needed to ensure a stable population. (This replacement figure has to be more than two children per married couple, owing to early death/infertility/unwillingness to marry/unwillingness to reproduce among a small proportion of the total community.) Therefore, without any deliberate governmental intervention, the people have achieved just the kind of drop in birth rate the Government would have wished to achieve. It is a remarkable event.

Britain, by comparison, was giving birth to an average family of 2.5 children in 1967 and this had dropped to 2.3 by 1972. As Britain, and the rest of Europe, are frequently belatedly subject (or subjected) to events, or things, or styles of existence first fashioned in America it may well be that

Britain and the continent will follow suit. What a boon it would be for Europe if this new birth-rate trend were also to come across the Atlantic with similar ease.

The only safe forecast is, as they say, that all forecasts are always wrong. Even in Britain, where the change in breeding behaviour has been less dramatic, the predictions have been see-sawing wildly. In 1955 our turn-of-the-century numbers for AD 2000 were expected to be fifty-three million. By 1960 they had shot up to sixty-four million; by 1965 up to seventy-six million; by 1970 down to sixty-six million, and every prediction since then has opted for a lower figure. In the United States everything has been more fiercely awry. The Census Bureau had to admit in March 1970 that every one of its four projections for the under-five age group had been wrong. Normally it produces four guesses for the future, ranging from high to low. When the famous revelation came out about the under-fives in 1970, even the lowest of the Bureau forecasts was far too high, estimating 2.5 children per young couple and not the 2.04 that had proved to be the case. Promptly, and unashamedly, the predictors produced a fifth projection based on the new evidence.

Confronted by the startling facts there has been no shortage of commentators with further facts all ready to account for the breeding revolution: a 'greater use of contraceptives' (65% of Americans were using them in 1970); 'higher abortion levels' (19% of American babies are still reckoned as 'unwanted'); more 'outside interests' for women; a 'reluctance to produce large families by those who were themselves members of large families'; a 'determination by youth to revolt against *all* prevailing customs'; a 'greater awareness of the uncertainties endemic in the world'. It is also feasible that continued talk about population explosions, ZPG, and starving millions has a cumulative effect, depressing the desire to have even one child. Or perhaps, as the young generation has already upset so many apple carts, it will more than compensate for its present apparent infertility by producing many more babies at a much-later-than-average time in its earlier-than-average marriages, thereby confounding the predictors once again.

On this score George Grier, of the Washington Center for Metropolitan Studies, has advocated in an arresting booklet called *The Baby Bust* a moratorium on all population projections, partly because we often believe them (and therefore use them as the basis for many important decisions involving much money, new schools, hospitals, etc.) and also because projection procedures could then be re-examined and improved.

If any country is going to have a population policy, and is not using Draconian measures to implement it, the predictions must be so good that all the accepted manipulations (taxes, benefits, home loans) can be correctly made in good time. And the predictions themselves must not confuse the picture by being responsible for amending the very event that they have – so skilfully – predicted.

Whatever the results of the present trends they will entirely influence the genetics of *Homo sapiens*. If this region breeds rapidly, and that does not; if these families limit their progeny and those do not; if this group is affected by governmental strictures, and that group is not; or if this individual reacts differently to his neighbour – all such differences will adjust the genetic material passed from one age to the next. Future generations can only build on the genes that earlier generations have handed down to them, and today's populations are behaving with no more regard for this fact than ever before although now, for the first time in the history of mankind, there is doubt about such capriciousness. Ought we to care more? And what on earth ought we to do? Before our current age, to shift the context of Enrico Fermi's famous words, 'we did not understand the subject; we still do not understand it, but on a higher level'.

We can worry about quality for the future, but in fact we cannot yet control our quantity. That, surely, is the first step. If America, and to a lesser extent Europe, are anything to go by, the people themselves seem to be making the decision. Without official stricture, without any international intervention, the sudden drop in family size caused by today's breeding group is extraordinary news. The explosion may have been tamed, at least in part, by nothing more and nothing less than a personal determination across the land to control it. 'Freedom,' said Hegel, 'is the recognition of necessity.' The generation more free than any other in history is proving his dictum and beyond all expectation. It has elected to carry out its own form of quantity control; it may even either now or soon be similarly forthright about some form of quality control. Who can hope to know?

CHAPTER 10

Death and accident

Premature infertility – Disease – New ills for old – Road
deaths – Wars – Domestic accidents – Suicide – Evolutionary
consequence

Eugenics is quality control. Under its banner, and in one form or another,
there will undoubtedly be steps taken in the future that will profit succeeding generations. However, so far as this book is concerned, that statement is premature. It is first necessary to examine three most particular
ways in which evolution is occurring. By no means, just because we no
longer live in caves or provide meat for sabre-toothed carnivores, has that
entirely straightforward process come to an end. Natural selection is still
firmly with us and is still operating as before – via the genetic material
handed from one generation to the next. It has no other way. As this is a
form of quality control, far more natural than any other system we may
eventually contrive, it takes precedence. Here therefore are some pages
dedicated to current evolution, to the ways and means in which genetic
material is *not* passed on to the future. Abortion is one method, contraception is another, but these two man-contrived manipulations must also
yield precedence to the natural process of death. To die prematurely, from
whatever cause, is to suffer a form of infertility. It is evolutionary failure
and it means, whenever it occurs, that a particular combination of genetic
material can no longer affect the next generation. It curtails a person's
breeding potential as abruptly as those carnivorous teeth and always will.
Today's numbers are different but the principle of selection is still just
the same as it ever was.

In the old days there was much premature death; early mortality was
the rule rather than the exception. Even today there are countries, for
example Algeria, Egypt and Guatemala, where more than half the offspring born alive then die before reaching the age of five. In West Africa,
where the expectation of life is the world's lowest, it is the process of
weaning that is largely to blame; the early regimen of reasonably sterile
breast milk is suddenly replaced by extremely unsterile and largely unsuit-

162

able foods. At all events the babies die, and might as well not have lived for all the genetic effect they will have upon the future of mankind.

The blessing of being able to survive smallpox, already mentioned, is of less benefit today. That kind of selection pressure is now being applied to those who can overcome the post-weaning gastro-enteritis of West Africa, or the man-made hazards of western society, or the animal-borne zoonoses that plague so many. Their effects shift year by year but every fact relating to death as a cause of infertility has implications for our future. So here, selected from a wide canvas, are some of them.

Smallpox, to elaborate further on this point, is a suitable starter. Mexico suffered $3\frac{1}{2}$ million deaths in the sixteenth century, for example, after the disease had been introduced from Europe, and there have been huge and recorded epidemics ever since man has been recording anything. Their lethal conquests only receded after Edward Jenner (in 1796) had discovered the principle of vaccination. In 1967 there were still $2\frac{1}{2}$ million cases but today the annual figure is approaching 50,000. The disease has been cleared from all but six countries – Britain became clear in 1935 – and annual deaths are now 15,000. Therefore the killer originally slaying 20% or more of us is now only destroying the life of one person in each 2,300. The current objective of the World Health Organization is to reduce even that proportion and rid the world, for once and for all time, of smallpox. Our inherited ability to resist the disease will therefore become a valueless blessing. Keep this point firmly in mind throughout this chapter.

Of course there were and are other ancient diseases. There are 15 to 20 million cases still of tuberculosis, predominantly in the developing countries but still powerful in the advanced communities where it causes as many deaths as all the other infections added together. The famous zoonoses, the 150 severe infections transmitted from animals to man, are not retreating everywhere but the general picture is one of declining influence. There is leptospirosis, the most widespread, which can be lessened wherever men and animals can be prevented from sharing the same water, as in rice fields. There is trypanosomiasis, still affecting 30% of the population in some African areas, but preventable wherever the tsetse fly can be prevented. Some, such as rabies, have justifiably horrific reputations but do not kill more than hundreds every year. Their names are powerful; the facts less so.

So too with cholera, a disease with no other host than man. Only on seven occasions in recent history has it broken out from the lands where it is endemic to become a pandemic elsewhere. It is quickly pounced upon, fairly easy to curtail but hardest to eradicate from areas where sanitation is deficient. It too is waning. A few of the old diseases are increasing, such as various food-borne infections owing to a greater centralization of food production and a disruption of old ways, but the gains everywhere have quite outclassed the ancient losses.

There is still influenza creating pandemics from some single focus, generally in the Far East, every decade and producing epidemics of lesser intensity about three times as frequently. Even though the virus is disturbingly able to change its character, namely its protein envelope components, it too is being forced to retreat, or rather it causes fewer casualties with every fresh advance. It is most triumphant over the old, the sick and the crowded, but medical care and good use of antibiotics to prevent harmful complications can reduce its power. To suffer flu today, and to survive, is of course genetically important, but the current infections cannot be compared, for example, with the pandemic of 1919. It showed that even the slaughter of war could be rivalled by the most persistent viral disease to affect mankind. Resistance to influenza, and the inheritance of that ability, has not been unimportant in our history: its form of natural selection has been extraordinarily powerful over the long years of its influence.

When the benefit goes, the resistance to a disease is no longer important; only the disadvantages remain, the genetic effects that almost certainly partner any advantageous mutation. A single benefit is unlikely to exist on its own because each change will probably have altered other attributes and these other alterations will, as with all mutations, probably be harmful. In time natural selection will be able to lessen their influence, but time has hardly been involved as yet. Britain only understood the cause of cholera six generations ago. Yellow fever was only traced to the mosquito at the turn of the century, just three generations ago. The war against malaria, that scourge of so many tropical countries, only began in earnest when World War II had ended, one solitary generation away from us. Mankind and his primate ancestors have been suffering from parasitic invasion, such as malaria, for many millions of years and for many tens of thousands of generations. No amount of blustering impatience on our part can cause the delicate balance of accumulated mutations to be immediately overturned and pushed aside merely because we have discovered DDT.

It is salutary to be reminded of the recentness of much biological advance. It so happens that the World Health Organization was initiated just one generation ago, in 1948. Infant mortality – first-year deaths – was not good even in those modern times. It has since changed: from 108 per thousand births in Poland to 34 per thousand; from 51 per thousand in France to 19.6; from 31.2 in Switzerland to 15.4. There will be further improvements because Sweden today, with the world's lowest figure, has only thirteen children dying in their first year for every thousand born. Other countries will catch up, or should, such as Pakistan, still with 130 dying, or Chile with a surprising 91.

What has also happened very recently has been the arrival of today's major afflictions. Soldiers returning from the Crimean war of 120 years ago are generally thought responsible for introducing the paper and tobacco cigarette to western Europe, and the resulting mortality has eclipsed over and over again the casualty lists from that war. There are almost 100,000 premature deaths annually in the United Kingdom alone, according to the Department of Health and Social Security, from smoking-induced diseases. One in eight of all deaths among men aged 35 to 44, one in four among men aged 45 to 64, and one in five of male deaths between 65 and 74, are said to be due to the cigarette. World-wide totals must be colossal, and more than sufficient to set beside some of those pandemics of the past. Tuberculosis deaths in Britain may have dropped from 21,000 in 1948 to 2,000 twenty years later, but lung cancer has risen during the same period from 10,000 to 29,000. Tit, in short, for tat.

It has been calculated, again for Britain, that 190,000 man-years of working life are lost annually from smoking, a figure omitting early death after sixty-five by which time men are presumed to have stopped work. It is a considerable total but, unlike many of the ancient killer diseases that hit young people hardest, smoking is mainly reducing life for the older man. Therefore any genetic predisposition to smoking, or any genetic resistance to its harmfulness, cannot so readily encounter selection pressures. The smoking male will, probably, have done his breeding by the time he dies from the carcinogens he had such pleasure in inhaling. Women, smoking less, using filters more and inhaling less deeply, seem to have a reduced constitutional (or genetic?) susceptibility to the malevolence of cigarettes. Smoking gives them smaller babies, by 7 ounces on average, but it does not kill the mothers, in the main, until they too are also past their breeding period. Once more, natural selection has nothing much into which to sink its teeth.

Accidents are quite different. These do affect the young. They must, in consequence, be affecting our evolution. People were killed during earlier centuries by the transport system, by horses bolting and shying, or kicking, trampling, and jettisoning their human load, but the numbers are not known. Today's epidemic, caused by the arrival of the motor car, is quite different. The first two road deaths in Britain occurred in 1896; and they were both pedestrians. Since then ever-increasing numbers of road victims – 250,000 dead this century in Britain alone – have been the occupants of cars, if only because cars are on the increase and pedestrians have become a rarer commodity. And since then, with the young becoming richer and cars more available, it is the young who have proved most vulnerable to the steel shaft of a steering wheel, to the combination of momentum and metal upon the human frame.

The link between young people and road deaths is most extreme in Germany – half of all the deaths among males aged 15–24 occur on its roads – but other European countries have associations only slightly less severe. To counter Germany's 50.2 %, there is Holland (with 47.3 % in the same age group), Denmark (45.6 %), Belgium (44.2 %), France (43.8 %), Austria (43.3 %), and England and Wales (41.9 %). In Poland and Hungary, where the young are less likely to have cars for themselves, the percentage is down to 14. Almost everywhere, as another fact not irrelevant to evolution, men are three or four or, in some countries, almost five times as likely to die in road accidents as are women.

The United States, with more cars per person, with great distances, and with more human time spent within its vehicles, leads the world in road deaths. That ticking clock in the US Department of Commerce may have indicated the arrival of the 200 millionth American citizen in 1967, but there ought to have been another machine to record the fact that, in November 1972, the one millionth American citizen died on its highways. World-wide, and with hideous regularity, one person dies every five minutes in a traffic accident, or 288 a day, or 8,600 a month, or 105,000 a year.

By comparison the Vietnam war, which lasted officially from, more or less, early 1961 to early 1973, killed 46,000 Americans, 183,000 South Vietnamese soldiers and (an American estimate) 925,000 North Vietnamese soldiers. During those twelve years 1,200,000 people were killed on the world's peaceful roads or precisely the number of military dead in this one aggressive war. In any single year more Americans die on their roads than died during those twelve years in the swamps and booby-traps of Indo-China. Wars also kill more young than old, more men than women, and the courageous more than their counterparts, but wars – or so the world hopes – will become increasingly a matter for history while accidents, or so it is easy to believe, will be with us for all time.

There are fatalities disconnected with the traffic holocaust that are also not impartial in striking this or that kind of human being. The person who falls off a ladder, or trips downstairs, or fails to take some elementary precaution, may appear to have been struck, blindly, by fate. But some kinds of people, and we all know them, are more likely to fall off anything. We say they are prone to accidents, but their proneness, for them and all their kind, often leads to death, to infertility from one more considerable cause.

In the United States (in 1968) 20 million people were injured in domestic accidents, with 110,000 suffering permanent injuries and 28,500 dying as a result. These millions either fell (40%) or were burnt (24%), poisoned (6%), shot (5%), electrocuted or hit (13%), gassed (4%), or asphyxiated (8%). Children under four and people over sixty-five suffer more accidents than all other age groups combined. One-year-olds are likely to swallow polishes, pesticides, petroleum products and dyes. The 2-year-old is more ready to ingest medicines, notably the omnipresent aspirins.

Once again the sexes do not figure equally in the facts. Most accidents happen in the home, and women are particularly subject to them. Women fall about twice as frequently as men, or at least they hurt themselves more so when they do. Of all a house's rooms the kitchen is the most lethal, and one definition of a developed society has been linked with the number of objects around the home which can be fatal. Children are most apt at detecting the potential weakness in any system, and 90% of all poisoning cases involve infants under five. Food poisoning, of which we all have a reasonable anxiety, is relatively harmless, killing in Britain only about 50 a year, as against 500 a year for the consumption of things never intended as food. Drowning, contrarily afflicting the swimmer more than the non-swimmer, kills twice or three times as many as do all those poisonings.

Such figures tend to be consistent, year by year, however accidental all the various causes. Each mishap is unpredictable, but the resultant total can be foretold with disarming accuracy. There is in consequence, for all of us, a degree of risk in going about our business, whatever it is. For a man, his wife and three children, the odds each year are 1 in 2,500 that one of them will be dead by the end of it from a domestic accident. There is a further collective risk, of 1 in 5,000, that one of them will die as a pedestrian. The man is likely to run an additional risk of 1 in 5,000 per year that he will die at work, the risk being much higher in certain industries, such as construction. As a car driver the man will carry a further annual risk of one in 5,000 that he will be killed while driving.

Society accepts such risks, particularly if they are of long standing, and yet they are all perpetually, and selectively, causing the line-up of human beings to change from year to year. Although old people are particularly prone to falls, and their age makes them irrelevant to evolution, the main cause of death in children over one is accident. This continues to be so until quite late in life and well after the breeding period. All youthful misadventure is therefore highly pertinent to evolution.

Fatal accidents, as the saying goes, only happen to other people. Yet, taken as a whole, and although the probability for men of dying by accident in western Europe is only 5 % (the range is from 7.2 % in Austria to 3.3 % in England and Wales), male life could become significantly longer if accidents were abolished. For women fatal accidents are everywhere less important. Their probability of becoming accidentally dead in Europe is 3 %, with the range varying from 5.3 % in France to 1 % in Malta. As men suffer more from this cause it is only fair that all figures in the succeeding few paragraphs apply only to them.

In accident-prone Austria, the current expectation of life at birth (for males) is 66.5 years. If infectious diseases could be eliminated this expectation would rise to 66.9 years. If accidents were prevented, and infectious diseases were retained, the life expectancy would be 68.5 years. Even if cancer were eliminated, and both accidents and infectious diseases were retained, life expectancy would rise only to 69.1 years. In less accident-prone communities the figures are not so dramatic but they still point an unwavering finger at the power of accidents to curtail the life (and breeding) span. Male citizens of Denmark can now expect to live for 70.6 years. Without cancer they would live for 2.8 more years; without infectious diseases for 1 more month; and without accidents for 1 more year and 2 more months. The same figures for England and Wales are 2.3 years, 0.2 years and 0.9 years, each of which could be added to the normal expectancy of 69 years if the causes were, somehow, eliminated.

In time, and when cancer has been brought down to the current level of mortality from infectious diseases, today's accident figures will become proportionately even more important. There are high hopes that all disease, including cancer, will be diminished in stature and its current virulence less triumphant. However, no one who has watched the steady progress of accidents, this fifth horseman of the Apocalypse, can have hopes much above ground level that its fatal advance will speedily slacken. To die young in the future will be, in effect, to die from an accident. To be infertile will be, increasingly, a matter of mishap and not of biochemistry.

As one more premature cause of dying, there is suicide, the deliberately inflicted accident leading to death. About a decade ago the figures in technological countries were roughly equal to the road toll. Recently, although the road damage has been consistent or worse, the British suicide figures have been improving. From about 7,000 a year in 1963, the total has now fallen to 4,000. Oddly this drop has not been followed by the rest of Europe or the United States, but the decline does not apply equally to every age group. A decade ago suicide numbers increased steadily with every age group and the current British phenomenon is helping to equalize the rate for the different groups. For example, in 1963 (in England and Wales) 39 men in every 100,000 over the age of 75 killed themselves, and this figure fell to 24 seven years later. In 1963 only 7 men out of 100,000 in the 15–24 age group were suicides and the figure fell to 6 seven years later. As a generalization, those below 45 years of age are now killing themselves slightly less frequently, those over that age much less so. With women the trend is the same although, in every age group, they kill themselves about half as frequently as the men. Both men and women over the age of 45 are, in the main, past the breeding period. Therefore, unfortunately, the current trend means that the suicide total is becoming relatively more important in evolutionary terms than in the past. It is one more form of infertility and, for the reproductive portion of the community, of considerable importance.

There has been repeated mention of the age at which death interrupts the normal span of life. For a person to die at 2 or 12 or 22 is quite different from 42 or 62 or 82. Natural selection is much influenced by the former and not, in general, by the latter. Harmful genes tend to be eliminated from the population if their bearers are less capable of producing offspring, whether because of straight infertility, or unwillingness to breed, or fatal accident, or death in any form. Natural selection becomes less efficient in purging deleterious genes (Peter Medawar's phrase) as individuals get older.

If only mankind continued to breed to the end of his and her days, as happens with many species, there could be a steady purging of unwelcome genes. Natural selection could then continue to operate. But, if a human dies at 50 or at 90, what kind of selection pressure can be brought to bear on the genes relating to his or her time of death when, in both cases, the genes have already been handed down? However, by worrying about that point, it is easier to understand the greater significance of early killers, such as most of the ancient diseases, gastro-enteritis, many kinds of accident and road deaths, as against the late killers, such as most cancers, smoking, heart disease. The first are a type of infertility; the second are not. The first are therefore extremely pertinent, time and time again, to various other sections of this book and their hidden influence must always be assumed.

CHAPTER 11

The infertility of contraception

Contradictions – Historical efforts – Modern procedures –
Free service – Governmental involvement – Vasectomy –
Future ideas – Sperm banks – Evolutionary significance –
Medical involvement – The right to prevent

*'Lorna; the Pill has to be taken by mouth. I think you should go
back to the clinic for further explanation of its use'*
Sunday Gleaner ADVICE COLUMN, JAMAICA

*'Male sterilisation is unacceptable if it is merely for allowing
sexual intercourse without the danger of conception'*
LORD DENNING (in 1954)

'If I don't have it done my friends will suspect the truth'
A VASECTOMY PATIENT

*'What is the good of going to bed early to save candles if the
result be twins'*
ANON.

'The rich get rich and the poor get children'
A SONG

The prevention of birth in any manner is highly relevant to mankind's
future, and for several reasons. It is extremely important what kind of
people are doing the preventing, and how effectively. It is also clear that
conscious decisions to amend the birth rate will alter ethical attitudes to
every other manipulation. The execution of abortion, more extensive now
than ever before, is having major effects in every direction, on numbers,
on opinion and on moral judgements. Finally, the artificial limitation of
each couple's offspring is, so many argue, a means of preventing evolution from creating the selection benefits to which we and all other species
have grown accustomed.

Today there is confusion on birth control, with great change and great reaction occurring simultaneously. Sweden sells condoms from slot machines, is good at sex education and liberal with abortion, but a Swedish man has to fly to London for a vasectomy. Japan is both backward and forward simultaneously. In the past forty years its birth rate has been reduced from 35 to 14 per 1,000 and the age range of mothers from 45 years to a remarkable 19. Nevertheless, according to Malcolm Potts (of the International Planned Parenthood Federation, to whom I am grateful for many of these facts), both IUDs and contraceptive pills are illegal in Japan, even though the Japanese 'make the thinnest, most beautiful, pastel-coloured condoms in the world' and perform almost a million abortions a year. The Pope has been condemned by many for 'archaic encyclicals', which fail to favour any control procedures save for the hazard of the safe period, and yet the birth rates and abortion figures for the 550 million Catholics grouped around the world are not noticeably different from those of neighbouring communities. South America, for example, weak in contraception, strong in religion, has the biggest abortion record of them all. Ireland, no less Catholic, has the lowest birth rate in Europe. 'Good Lord, you can procreate without having babies,' quoted the Paris *Herald Tribune* after overhearing this Irish conversation.

France, with an estimated 50% of all conceptions ending in abortion, did not permit the pill to be prescribed until 1969, nine years after it had become generally available. The first Italian birth control clinic of all time was not opened until that same year of 1969. The first such British clinic had been opened in London in 1921, but in 1972 London Transport steadfastly refused an advertisement for a family planning clinic. The Independent Broadcasting Authority of Britain was only initiated as an organization in 1955 but began with a code prohibiting the advertising on television of anything to do with birth control. In 1970 this was amended to permit advertisements from family planning advisory services but not about contraceptives.

Such contradictions exist but their erosion is on the increase. The number of Family Planning Association clinics in Britain has been growing by about a hundred a year, as many therefore since 1966 as during the first forty-five years since Marie Stopes had the temerity to open up in Holloway. More and more people are realizing the obvious truth that the prevention of unwanted children also means the reduction of equally unwanted population pressure. Americans, in assessing the double potential blessing, have reckoned that their undesired pregnancies may add up to 30% of the total.

Many countries have produced similar results. Take England and Wales in 1971. There were 783,155 live births, of which 65,678 were illegitimate. A small proportion of these babies, created by what the Finns call, more reasonably, the union of 'illegitimate parents', must have been wanted, say an eighth. In the same year there were 123,091 legal abortions with, presumably, a far smaller number of these conceptions being planned. Surveys have also shown that about one-third of all pregnancies within marriage are unplanned, a large number of which were taken care of by the 55,540 abortions (in 1971) for married women. Therefore, by adding up both the facts and the assumptions, the total of unplanned pregnancies in one country – England and Wales – during that single year was over a third of a million.

Of course many unwanted pregnancies become exceptionally wanted children, but that is another story. The bare fact remains that in 1971, in the third most congested country in the world which is confronted by its density at every turn, a total of less than 500,000 pregnancies were actually planned. During the same twelve months, and at the other end of the spectrum, a total of 567,000 people died. Therefore, deaths would have greatly exceeded births had only the planned pregnancies been those to achieve full term. The actual, and unplanned, picture was quite different: births substantially exceeded deaths, despite the thousands of abortions, despite the national wish to reduce the population pressure and the fact that English people are no more idiotic than many others in managing their own affairs. Nevertheless any stranger to our planet might think the facts exceeding odd.

The story is not new. There have been unwelcome pregnancies for millennia, just as there have been attempts to control them. Contraception ideas have even been inscribed for some 4,000 years, certainly since the Petrie papyrus of 1850 BC which recommended crocodile's dung and honey. The variety of schemes exercised since then, cooked menses, railway trains passing overhead, post-coital gymnastics, wolf's testicles, says much for mankind's original thinking but little for the ability to solve the situation. The condom arrived by name at the beginning of the eighteenth century, but trust in the various sheaths under this heading was never so certain as their aptitude to reduce the requisite sensation, the whole cause of the problem. With the arrival of new rubber techniques in the twentieth century, and a capability to make sheaths both suitably thin and impervious to sperm, the age of contraception can be said to have begun. (The London Rubber Company, current giant in a business combining toy balloons and rubber sheaths, began in 1916 in a small back room. British production is about 2½ million sheaths a day and somewhat fewer balloons.)

The rings, coils, loops and other devices, collectively known as IUDs, and about which there is considerable uncertainty – not to say complete ignorance – concerning their mode of operation, originated in the agitated Berlin of the 1920s, but only received wide application in the last two decades. The various diaphragms have had a longer history, having their modern origin in early nineteenth-century Germany, and the spermicidal extras have been gradually improved. Nevertheless, however gossamer the sheath, or unobtrusive the IUD and other devices stowed away within much of womankind, there is something undeniably primitive about these barriers to conception. Any libertine from the past would only have needed a few moments' instruction to feel warmly at home in modern times. 'Contraceptive methods are so crude,' said Alan Parkes, eugenicist, biologist, 'as to disgrace science in this age of technical achievement.'

Then came the pill, the first scientific contraceptive, the first to make use of hormonal and glandular knowledge. (It should correctly be called a tablet, but momentum is too strong for that particular change.) First mentioned internationally in 1955, first tried experimentally in America and first given large-scale tests in Puerto Rico soon afterwards, the pill was poised for general use by 1960. Americans were quickest off the mark, with 24% of all married women who were using any form of contraceptive taking the pill even by 1965 and 34% by 1970. This reaction was an 'amazing phenomenon', said Charles F. Westoff, co-director of the 1970 National Fertility Study. Britain was initially more hesitant. In 1962 there were 3,500, cautiously, on the pill. By 1964 – 44,000; by 1968 – a million; and by 1972 – two million. The pill price, everywhere, has been falling year by year as the business has grown. Medieval libertines would not understand the little tabloid object, it being entirely dissociated from the engineering of dams and barriers. They would be astonished by it and would know that a new age of contraception had truly come to pass.

The dark ages of crocodile dung may therefore be over, but they might not be if judged by the number of people who still seem to disregard the need for contraception. Hammering home this point are the thousands who request abortions – the total for England and Wales was 156,000 in

1972 and greater still in 1973. Only a fifth of the British women 'at risk' are taking the pill and in most other countries it is very much less. Already there have been protests, and not only from the Church, both at the high rate of abortions and at the numbers of unwanted children actually being born when there is good contraception – and safe abortion – to prevent them. Therefore it is certain that various mandatory schemes, aimed at making more sense out of a senseless situation, will increasingly be advocated.

Melvin M. Ketchel, physiology professor at Tufts Medical School, suggested in 1968 that a safe fertility control drug might be added to the water supply. Any such measure, he felt, was preferable to the starvation and misery resulting from overpopulation. Others have proposed that all conception should be made a positive step, rather than a lack of the positive step of contraception. Hormones suitably encapsulated, and placed beneath the skin at puberty, would have to be removed whenever conception was desired.

Coupled with this scheme could be the birth-ration system, which has caused much shaking of heads at the very thought of it. Assume for the moment that a powerful government had decided the necessary replacement level was 2.2 children per family. Each couple, either married or with some stated assertion that they wished to breed, would then be allocated two cards. The 0.2 spare child allowance could be granted, first come first served, to all those wanting a third child. The number of successful applicants could, of course, total not more than 1 in 5 of the original couples. Should any couple change their mind, or prove to be infertile, they would be permitted a measure of choice either to return their cards to the state for further allocation or to donate them to friends who wanted more children. And presumably the things could also be bought and sold.

The system does not have instant appeal to all, even though in communities of little growth the end result would not be greatly different from the current situation. In countries like India, where traditions are somewhat inflexible, and where most people live their entire lives without the single attendance of a medical practitioner, one can imagine the fervent disdain with which a village might greet the card distributor. Societies forcibly prevented from such disdain are also imaginable, and at roughly this point the idea of birth-ration cards can lose every particle of its simplistic appeal.

A far better scheme, or so it is generally felt, would be for individual choice to be maintained but for the State to promote every opportunity that might reduce family size. April 1, 1974 was the date on which Britain took a crucial step forward. On that day the National Health Service was reorganized: family planning clinics, including those operated by local authorities, were brought under the wing of the NHS. On that day the malaise of heedless childbirth was brought into line with every other illness. It became the direct responsibility, via the NHS, of the government. It was – almost – the end of a long campaign, partly assisted by the earlier abortion law that had turned doctors ethically upside-down while they expunged young foetal lives at the rate of hundreds every day. They had generally conceded that any plan to oppose such a deluge was plainly preferable to this slaughter.

To begin with, and in 1972, the bill presented to the House of Commons had been for free contraceptives and advice to be awarded to all women recently delivered of either a baby or an abortion. This curious logic, providing stable doors only if a horse had already departed, was less acceptable to the House of Lords. Britain's second enclave of governmental opinion then recommended free examination, free treatment and free contraceptives. Finally, following a dramatic change of heart, Keith Joseph (Secretary of State for Social Services) announced to the Commons that contraceptives should be free for all.

There were anomalies. Much discussion was given to the problem of prescription charges. It was argued that, as with drugs, these additional sums (paid for each new prescription) should also be paid when acquiring each new consignment of contraceptives. The cost of such extra charges for the average individual would be about 80 pence ($2) a year, creating the bizarre situation that abortion would be free while contraception would be (modestly) taxed. Certain people are exempt from these prescription payments, such as those under 15, over 65 or expectant, an odd group to benefit from a reduced fee for contraception. However, when April 1 approached, it became clear that the principal stumbling block concerned the medical profession itself. There was talk that doctors should aid the sick and not consume valuable time assisting the licentious to indulge without anxiety. As yet, although general practitioners prescribe contraceptives for those felt to be needy, they are not yet participating fully in the new and dramatic shift towards free contraception for all.

As dramatic changes go it will be cheap, even when everyone is involved in the new situation. The annual bill of Britain's National Health Service is £2,700 million, and the addition of free contraception to that enormous figure will be about £17 million. Money spent in preventing births is undoubtedly an investment at the present time, and counters Churchill's famous statement of 1943 that 'there is no finer investment for any community than putting milk into babies'. Without doubt it reduces the bills for health, welfare and education. Lord Gardiner, judge and spokesman in the British House of Lords, has said that each £1 spent on family planning will save the taxpayer about £100. Other estimates have declared that every illegitimate child costs over £4,000 in supplementary benefits; therefore prevention of all such children would bring a commensurate return. The city of Aberdeen in Scotland went out on a limb in 1967, spent $2\frac{1}{2}$% of the local authority budget on a free family planning service, and is already reaping rewards. It has achieved extraordinarily low infant mortality, a big drop in the number of large families, a population growth of 0.27% (meaning a doubling every 400 years) and world renown. Already the number of pupils per teacher, for example, has dropped from 23.6 in 1967 to 21.2 within five years. To a teacher the gradual removal of 2.4 children may not seem a staggering or even detectable event, but to the planners, the futurists and the city council all such increments are exciting, rewarding and a change for the better.

To India, or the world in general, such changes would be a revolution in themselves. If Britain's cost of an entirely free contraceptive service is typical – £17 million a year for 50 million people – the actual cost of providing the information, service and appliances would not be colossal: £176 million annually for India or a total of £814 million for the entire world. Such a figure is much less than the arms budget of many countries (it is a fifth of Britain's military expenditure) and such a sum may one day be spent on this more appealing form of people control.

The prospect of two children per couple, of a population merely replacing itself, does not yet have universal acceptance. There are those, for example, who feel it to be a negation of evolution. 'When will it be realized,' wrote Brian Benham to *New Scientist*, 'that population stabilization due to birth control must only result in the end of the human race as we know it?' Basically this argument is that more offspring must be produced than can survive, that selection against the unfit must be permitted to take place and such a culling is necessary if mankind, or any other species, is not to deteriorate. A system where every child is planned, and is then born into an environment able to ensure its survival, is not one – so the argument runs – in which the proven method of natural selection can readily operate. Today's approach, whereby the physically or mentally impaired have virtually as good a chance of survival and even of breeding as the others, might lead in time to the greater inhumanity of a weakened species.

As a counter to that argument is the fact that the very artificiality now in existence has led to some striking and beneficial blessings. In Japan the figures are most dramatic. The great birth-rate drop there since World War II has been accompanied by reductions in congenital defects. According to Ei Matsunaga (of Yata's Department of Human Genetics) this bonus was due 'to the decreasing frequencies of births of higher rank mothers as well as those of both older and very young mothers'. From 1947 to 1960 the subsequent reduction in defects was about one-third for mongolism, slightly less for other chromosomal errors (XXY and XXX), one-half for the unwelcome Rh-erythroblastosis, and about one-tenth for all remaining congenital defects. Against these undoubted eugenic gains are the dysgenic effects of preserving harmful genes; but smaller families, should they continue, and a lowered practice of inbreeding will mean that the manifestation of their harmful capabilities is less immediately likely.

Contraception, in short, is welcomed, disdained, expensive for some, cheap for the state, a manipulation of evolution, a kind of signpost for the years ahead and always with arresting points to make. Take, for instance, the very obvious requirement of a morning-after pill, coupled with the traditional human reluctance to wait for the right kind of medical assurances. Take, for example, the recent story of diethylstilboestrol, the synthetic oestrogen which has been used in the treatment of various gynaecological problems for over thirty years.

The hormone's modest fame was abruptly altered in October 1971 when an article appeared in the *Journal of the American Medical Association*. It described the drug's action as a post-coital pill in tests on 1,000 women. No pregnancy was reported from the trial even when the drug had been taken as long as seventy-two hours after coitus, and the printed facts were eagerly read on every campus. Women were able to obtain the drug fairly easily and, by the time a Senate committee was hearing about it in March 1973, the taking of DES had reached 'epidemic proportions'. America's enthusiastic predilection for the latest product, particularly when coupled with such urgent need, meant that over two million women had taken advantage of the hormone's properties within two years and 50% of campuses had been involved in its bush-fire distribution.

The wretched flaw to this involvement is that a clear link has been shown to exist between the administration of the drug to pregnant women and the development of vaginal or cervical cancer some 20 to 25 years later in any daughters born from those pregnancies. As with thalidomide the blow falls only upon the next generation. With DES there has been such a carcinogenic effect on tested animals that its extensive use in poultry rearing has even created anxiety in veterinary circles. Plainly the drug had slipped past rather than through the fine-mesh net of safety regulations created by the Food and Drug Administration of the US (which, to its eternal credit, prevented the distribution of thalidomide).

Once again the signs are clear for us to read. The campus behaviour indicates that a post-coital drug has a greater potential even than the pill – there were not two million taking that benefit within two years of its inception, but it is of course desperately to be hoped that the impulsiveness of 1971–3 will not be resented, agonizingly, by a great group of girls around the year 2001. DES is said to have worked well in fulfilling its post-coital need, but what about the few pregnancies that may, despite the drug's interference, have continued to term? Will the females among those foetuses be subject to the early cervical cancer that today's information seems to indicate? Thalidomide caused 7,000 babies to be malformed, but it was hailed originally as a blessing, being cheap, harmless to the mother and a great reducer of the nausea of pregnancy. DES was welcomed still more dramatically and the end of its story cannot yet be written.

Contraception may well have advanced from a primeval past, but plainly the system is still primitive. To have such a need as DES has exposed, and to have such a risk as well, shows up a fearful backwardness in our capabilities as against our desires. To have abortions as a massive means of preventing the millions of unwanted infants from being yet more numerous will, if the future looks back at our present, be viewed with bewilderment. It is easy to mock at those pessaries of crocodile dung in ancient Egypt, but current contradictions in the manner whereby we fail to limit, rather than actively promote, the generation to succeed us will surely be thought no less strange and certainly reprehensible.

Less reprehensible has been the sudden determination of many men to have done with their role in the possibility of conception as effectively as they – and modern techniques – can permit. Vasectomy, the permanent form of male contraception, has always been available but has only been used by large numbers of men in recent years. Letters to the *British Medical Journal* have discussed some of the procedures involved and they can make disturbing reading for any man whose ego is somehow wrapped up in his testicles.

'The vas must first be digitally fixed subcutaneously in the scrotum (a knack which can be learned and taught), a skin bleb raised, and then 2% lignocaine infiltrated into the sheath of the vas with a 5/8ths inch (16 mm) subcutaneous needle. A total of 10 ml of local anaesthetic or less is all that is required.'

'Difficulty in locating the vas is a matter of experience, and experience can be obtained before attempting vasectomy. Every male doctor has two vasa which he can spend a few minutes locating every day.'

'It is of course essential to use suitable instruments (Soonawalla forceps) to hold the vas once it is revealed through the incision, and when gripped by small curved mosquito artery forceps, which crush it before section, the ends cannot slip back and disappear. It is not necessary to ligate the cut ends, which can be sealed adequately by unipolar diathermy with a Birtcher Hyfrecator.'

'A few patients are tense, but this can be assessed at the preoperative discussion and can be dealt with by giving diazepam 10 mg orally half an hour preoperatively or, in rare cases, by giving 7–10 mg intravenously.'

'The only discomfort the patient experiences is on introduction of the local anaesthetic.'

'Except in one or two isolated cases the patients express surprise that the operation has been so painless, and immediately afterwards are able to walk or drive home unaccompanied.'

'A common remark is that it is less of an ordeal than "going to the dentist".'

Essentially what is happening is a severance of the pipe line existing between the sperm production area (the testes) and the storage area (the epididymis). It is not castration. It does not prevent ejaculation, for the bulk of each ejaculate is not sperm but subsequent additions. It should not decrease libido, for that is linked to the production of the hormone testosterone and not to the production of sperm. The operation is, as the letters above indicate, an out-patient procedure and will have to be if operating theatres are not to be swamped with this surgical form of contraception.

The technique relies upon the helpful fact that, for part of its rambling length, the vas deferens runs inside the scrotum and can readily be approached through the scrotal wall. (As with that quotation, every male – and not just doctors – can search for his vasa should he so wish.) Anaesthetic can be local, for most people, or even general for the unduly apprehensive, to prevent pain. A cut is made through the scrotal wall, the very thin tube is raised and a piece of it is cut out, perhaps 1 cm, perhaps 2 cm or even more. The two ends may then be bent over before being tied off, a method intended to reduce the chances of both tubes unhelpfully reuniting with each other later on. As there are two vasa the operation has to be repeated on the other side, but the whole procedure should not take longer than 15–20 minutes.

However short the visit to the doctor, however simple the operation, vasectomy is not immediately effective as a form of contraception. There are all those sperm, still unejaculated, that were on the penis side of the point of severance at the time of the operation. Not for twenty ejaculations or so – some clinics advise three months – can a man be fairly sure that he has exhausted his supply and is effectively infertile. The simplest test to give the necessary peace of mind and prove the operation's success is for an ejaculation to be examined under the microscope at some later date. Such sperm-counts are a routine procedure, but they can be yet more critical should a man's wife become pregnant long after he himself has been vasectomized. Was it a faulty operation or another man? Or was it recanalization, the disarming but cunning method whereby sperm are once again transported across the gap created in the clinic?

Medico-legally the ability to eject even one sperm is important even though, to anyone working in the infertility field, a sperm count below twenty million per millilitre is considered both abnormal and a probable cause of failure to conceive. In a letter to *The Lancet* a correspondent wrote (in May 1973): 'Colonel Walter states categorically that it is a legal fiction that a single sperm can cause pregnancy. He may be right, but how does he know?' A nice point. In fact, there have been definite cases of conception after vasectomy, but such events are rare. Cutting the vas can be considered, despite the hit or miss nature of much of biology, an effective method of contraception.

Therefore is it safe? Is there a price to be paid, such as physical side effects or other unforeseeable consequences, for the convenience? Here is one assessment: 'Vasectomy, and its effects upon the mammalian testis, have been discussed for more than half a century with approximately an equal amount of contention that degeneration follows occlusion of the vas deferens and that it does not'. This analysis was written back in 1924 and means, therefore, that the issue has been warmly debated for over a hundred years.

There is still no secure answer but time will quickly tell. With the rapidity of all recent events in the contraceptive field, the cutting of the vas has at last come into its own. As with the pill, America is leading the world with 750,000 male operations in 1971, almost a million the following year, and another million in 1973. Britain, despite the creation of the Vasectomy Advancement Society (oh, what ingenuity) in 1971, has only performed a total of some 50,000 operations, but is still ahead of any other European country. It will soon go further ahead because vasectomy will shortly be available – free – on the National Health Service along with other methods of contraception, a policy described as 'mutilation on the rates' by its detractors. India is the major developing/poor nation to have put energy behind this manner of contraception, and has successfully severed the vasa of 12.5 million men in recent years. Unfortunately the procreative energy of that sub-continent is still winning; it produces more than that number of new citizens each and every year. Nevertheless the harmfulness, or otherwise, of the operation will quickly become manifest, now that millions of different categories of people no longer have spermatozoa flowing up their epididymis, behind their bladder, past their seminal vesicles, through their prostate gland and along their urethra to prove such a hazard to human welfare when finally dispatched from the penis.

Side effects may arise as counter-weight to the considerable blessing involved and disquieting accounts about vasectomized rats have already occurred. For example, four New York scientists writing in *Science* (in January 1973) reported that immature rats suffered a significant shrinkage of their testes and, more disturbingly, of their testosterone production – about 20 % less – following the vasa operation. A major difference, apart from the fact that rats and humans are different species, was that the rats were young whereas human volunteers for vasectomy are nearly always mature, having probably sired the number of children that they, and their wives, desire. Further experiments on more middle-aged rats are therefore important but, even so, the differences between rats and men might still confound the picture. A man, for example, even though experiencing the diminished effect of a smaller supply of masculine hormone, may be so relieved at the impossibility of conception that his libido and performance increase substantially above the normal level. Men are not rats, but in the absence of experiments upon men it is necessary to exploit laboratory animals until such time, not so long now, that the information from vasectomized men is sufficiently clear.

A side-effect no one had predicted, or rather no one had expected to see so rapidly, was the near-instant commercialization of several of the issues involved. There are now sperm banks. Vasectomy is not always reversible and, however much it is also not castration, the operation can give men doubts as they accept its implications. Are they quite sure they will never want to father another child? Might it not be better to bank a few of their sperm against the procreative equivalent of a rainy day? America, once again, has been first in the field. The Idant Corporation (of New York) and Genetics Laboratories Inc. (of Minneapolis), the two largest operators in this new line of business, both offer services to patrons wishing to establish this latest form of desirable credit.

The procedure is simple, the cost inexcessive. First refrain from intercourse to build up a certain capital supply. Then bring two or three instalments of this most personal of all deposits, preferably within two hours of its ejaculation. If such immediate errands are inconvenient, the postal service being out of the question, it is possible to attend the bank personally, to be shown into the ejaculatorium – their word for it, not mine – and to produce the necessary donations on the spot. A certain amount of pornographic material is present to provide the necessary stimulation. One inevitably wonders how long it will be before the most effective stimulants of them all will replace the magazines, and will be lounging about decoratively for the customers to make use of this friendlier female way of being relieved of their contributions.

The cost of opening such an account is currently about £35, with a further £7 being the annual storage fee for keeping the sperm phials in liquid nitrogen at minus 196° C. This cost can be offset by supplying a greater quantity of semen – at £8 a time – with the understanding that the excess will be used for infertile couples in need of both artificial insemination and someone else's sperm. The business is booming. Idant started in 1971, it had branches in twenty American cities by the end of 1972, and was looking outward to Europe and Japan by 1973. There is more about the procedural complexities of insemination in Chapter 16, but the sperm banks will undoubtedly hasten decisions on many issues involved. Will these banks, for example, by catering initially for the reasonable doubts of men about to be vasectomized, gradually take over the business of storing sperm? Will they become eugenic as well as commercial in their thinking? Will they pay more for the sperm of brilliant or good-looking men? Will they, in contrary manner to those famous cash accounts of Switzerland, advertise their clients, enabling them to make a handy living through their ejaculations?

Also, what about security? Animal sperm has been kept for many years, and has been proved viable over long periods, but much less is known about human germ cells. Will any depositor have redress if his frozen assets prove to be worth no more than the phials they are stored in? Dr Alan Guttmacher, when president of Planned Parenthood, said that a man is deluding himself if he thinks he can come back a few years later, collect his sample, and then be sure of having children. Can he even be sure of collecting his own sperm, however steadfast the bank in asserting accuracy? At present there is scarcely any governmental legislation covering this kind of operation; but vasectomy has arrived, the banks have arrived and, as always, the rules will only arrive when injustice is not only done but seen to be done.

If the banks are legislated out of business something similar will assuredly take their place. Once again biological advance has permitted a need to occur. A man in Minnesota has banked his spermatozoa just in case his only son either will not or cannot carry on the family line. A crew man on a nuclear submarine, aware of the possibility of harmful radiation, has deposited his germ cells – just in case. Other men, perhaps believing that posterity will give them their due although the current generation has done nothing of the kind, may well feel – at $20 a year – the future merits a specimen of their innate capabilities. After all, people are already leaving their dead and frozen bodies for the future to revitalize; mere sperm not only cost less but are more likely to spring to action than diseased and aged corpses. Society itself, more aware than ever before of future generations and of conserving for them a decent measure of the planet's amenities, may well feel that selected sperm should also be preserved. Picasso is dead, and every care is being taken to cherish his works, his homes, his memories. It is not unthinkable that future Picassos will also leave behind that part of their mortal remains most susceptible to immortality.

It is not implausible, but may well turn out to be impossible. For some reason human sperm does not take kindly to lengthy storage. Artificial insemination is much more successful, namely pregnancies happen more surely, if fresh semen is used. If chilled and then warmed up again the spermatozoa are still satisfactory, but less so. If chilled for longer periods there is a further loss; more and more inseminations are required to make pregnancy occur. No more malformations are created, but something vital is lost with the passage of time in a frozen state. Bull semen is different and lasts extremely well in these circumstances. Pig semen is different yet again, being sharply intolerant of the treatment. Mankind may well prove to be nearer the boar than the bull despite every promotion of technique. And therefore mankind may prove to be incapable of this form of delayed fertility.

A brief word about the evolutionary consequences of contraception provides a timely interjection. Anyone who prevents a conception, whether by abstinence, or by stopping sperm from uniting with an egg, or by evicting a conceived but unwanted foetus, is preventing his or her gametes from reaching the next generation. He or she may have children already, but every attempt at contraception is an attempt at reducing one's genetic influence upon the future. The man who severs his vasa or the woman who renders herself equally infertile is helping to create an artificial selection. The next generation will therefore carry fewer genes from those people most adept at contraception, most willing to reduce their progeny, most firm at halting their own inheritance. Should such traits be heritable there will be less of them in the future, less willingness to refrain from breeding, less enthusiasm to limit family size. The effect may be impossible to measure, for the new environments of future centuries may swamp the inherited differences, but that is not to deny this difference. Evolution, to say it again, can only work through genetic transmission, and people who cancel their potential influence upon the future are leaving the field to others. Contraception is a form of infertility, a form of selection, and as such is most pertinent to the future of mankind, whether for good or ill.

Given the susceptibility of spermatozoa it is strange that more sophisticated methods have not been used to stop them. However gossamer the sheath, this device has about as much subtlety as the contraceptive pebble pushed by any Arab up a camel's vagina. Similarly, however effective and welcome vasectomy may be, it could have been achieved in Neolithic times with any sharpened flint. Chemical castration, some male equivalent of the pill, has not yet arrived, but at last may be on the way.

For at least twenty years there have been compounds which impair sperm production, but there have always been disadvantages to them. Usually they have been just too poisonous. If not poisonous their technique has been more like a sledge-hammer, with effects in every direction.

An example is cyproterone acetate. This is a steroid which counteracts male hormones without its effect being accompanied by the general feminizing that usually partners such counteraction. So far, therefore, so good.

However, its other defects are somewhat overwhelming. In blocking testosterone it reduces libido, it lessens the male capacity for an erection and it hinders orgasm. According to a research team more interested in finding blessings than drawbacks it also 'creates a sexual calm which can be a starting point for psychotherapy'. One company making the drug has expressed its awareness that there is only a limited market for this kind of pharmaceutical advance, with that market composed of a small number of unfortunates. For example, a man of sixty-one charged at Cardiff Assizes (in January 1973) with offences of indecency and attempted buggery, and with a long history of sexual offences, was placed on probation provided he attended the local psychiatric clinic. He did so, he was given the drug, he could no longer achieve an erection, and he was then said to be harmless sexually.

There should be more hope, and less cause for anxiety of every kind, in a technique devised by Dennis Lacy (of St Bartholomew's Hospital Medical College, London). It makes use of the fact that sperm production only occurs because the testes are stimulated by hormones from the pituitary, the ductless gland at the base of the brain. Shut down the pituitary, therefore, and the testes are also halted. Unfortunately, this gland's closure would halt virtually every aspect of a man's maleness. Therefore, so far so bad. However, according to Lacy, it is possible to stimulate everything else without stimulating the testes by means of a pill containing mesterolone coupled with progestogen, a substance also found, confusingly to an outsider, in some contraceptive pills for women. Experiments are now being conducted. Time, once again, will tell.

Should the technique succeed, and should it become popular, the onus will once again fall more heavily on the male to prevent conception. It was up to him in the old days of coitus interruptus, and it remains so whenever this is practised. It was up to him, and still is, with the condom, the most widely-used contraceptive of all time. Recently, with the pill and the IUD, the burden has fallen more heavily upon his consort. To some extent males are again taking over responsibility by offering themselves for vasectomy but women are also being sterilized. Of the 86,000 legal abortions in Britain in 1969 over 14,000 were followed by sterilization of the woman concerned. However, the relative simplicity of the male severance will keep the man ahead on this score.

Perhaps, say some, all energy on contraceptive research should be directed at one sex only. What is the point of arresting the mechanism by more than one method if one is sufficient? Why, in effect, remove both the rotor arm and the battery from a car when the absence of either will halt the system? At the moment, like some belligerent mechanic determined to throw a spanner in every portion of the works, we are attempting to stop sperm production, stop sperm access to the penis, halt its flow from the penis, stop its passage to the vagina, prevent its union with the egg, prevent that egg from implanting itself should it be fertilized, and then abort the whole lot should every one of these precautions have failed along the line.

The reason is that no kind of prevention is yet satisfactory. The endocrine insult of the pill is extreme; its links with high blood pressure unfortunate. IUDs can be troublesome to both women and their mates; they can be difficult to place, difficult to secure and can cause bleeding even when in place. The post-coital methods, however desirable in theory, are less acceptable in practice. Abortion at any stage is easy to decry. And the kinds of safeguard, of lengthy trials, necessary before any new oral contraceptive can come on to the market are certain to delay even the best of ideas, however rich the potential pickings for the developers. Nevertheless, this multilinear approach, and the tinkering at every stage of the conceptual process, add up to further manipulation of the reproductive act. Our attempts at contraception will surely lead to greater deliberation in the matter of conception itself. The decisions are being left, less to chance, more to ourselves. They are for us, increasingly, to control.

To what extent, therefore, is the medical profession going to find itself non-plussed by its new role? Already there is confusion. There is resentment at having to pump out foetuses. There is the desire to keep the pill on prescription but not the wish to prescribe condoms, yet both come under medical care with the new laws in Britain. There is also a prevailing sentiment that medicine is aimed at alleviation of illness and should not be concerned with various boons affecting our way of life. 'The vast majority of vasectomies carried out at present are purely convenience operations,' complained one doctor to *The Lancet* (in December 1972) before continuing: 'Surgeons are not merely technicians to be directed to do operations when it suits the patients. I have refused patients on the grounds of youth or the fact that they do not have what I call a minimal family, i.e. 2 children.'

Such convenience operations may be disliked but they can be the lesser evil. 'It has been my practice since the Abortion Act was passed to accede to almost all requests for sterilization lest, if I refuse, the refusal will sooner or later be followed by a request for termination,' wrote another doctor, again to *The Lancet*. Therefore at what stage is it right to put paid, at her demand, to a woman's fertility? 'Requests from younger women are often too capricious to be taken seriously . . .' 'It might be far wiser to restrict "social" sterilization to women over 35, though clearly there will be exceptional circumstances . . .'

A person's attendance at the local GP surgery can be unembarrassing for wholesome complaints, such as warts and rheumatics, but sexual safeguards are in a different category. The person's wife may be known to be pregnant or even infertile while the man is awkwardly asking their family doctor for contraceptive help. A girl's father may have expressed a desire for grandchildren while his childless daughter is asking for sterilization. A shy girl may be fearful of the kind of questioning that may be delivered along with her three months' pill supply. Even family planning clinics, happy to give advice to all, sometimes require a letter of introduction from a doctor, that same family doctor so frequently both judge and jury with the best of intentions.

Contraception is not used as extensively as might be imagined or as it should be used. Perhaps the doctors, with their moral concern, are partly to blame. Perhaps there is still too much difficulty in asking sexual questions, save from similarly inexperienced contemporaries. No contraceptive being sold has instructions on the packet and girls have been discovered with condoms inserted into their vaginas. In one Aberdeen survey (of 684 unmarried girls, all students at the university) 86% had a steady sexual relationship but no contraceptive procedure was practised by 30 to 40% of these couples. Why not, when an unwanted pregnancy can be a disaster for academic ambitions and certainly inconvenient? Six of the girls were pregnant at the time of the survey, forty-nine thought they might be and sixty-five had been pregnant in the past, with all but ten having had an abortion. If students are so casual, or uninformed, or fearful of 'moral interrogation' (their phrase), it is disturbing to speculate about their less privileged contemporaries.

The incidence of unwanted pregnancy is a major problem for doctors, said the *British Medical Journal* editorial discussing the Aberdeen survey, 'so long as doctors quite rightly regard the giving of advice on contraception and sexual activity as part of their job'. Perhaps the advice is less welcome than the contraceptives. Anyway, the survey itself was concluded with an evaluation: 'A doctor has the right to express his own attitudes and a duty to acquaint his patient with the relative risks of having intercourse, of pregnancy, and of the various methods of contraception, but if the patient, with these insights, wants a contraceptive the doctor should see that she is provided with the appropriate form available.' And so too, presumably, the male student who is her mate.

So who will call the tune? There is now abortion in many countries virtually on demand, and there is resentment. There is now a wide range of available contraception, free for all in certain countries, and once again there is a measure of resentment. So long as the medical profession is involved with the prevention of reproduction there will be greater safety – the right drugs will not be given to the wrong people – but there will be strictures along with the supply.

It would seem, when future change is contemplated, and when yet more forms of positive reproductive control become available, that the medical profession will have a still greater weight to bear. A few years back there was no vasectomy to speak of, not much female sterilization, less abortion (at least within the legal channels) and no pill. Much more must be ahead of us and possibly more resentment from practitioners at procedures demanded by a more aggressive public. Abortion did not come about by medical demand but medical men have to deal with it. The dilemma, with a greater control by ordinary people of their own reproductive capabilities set against a greater imposition on the medical profession, is and has been acute. Abortion's particular problems are unique, but this form of interference with normal reproduction can hint at future problems when interference is more rampant than today. The subject is a huge one, the implications are enormous. Most certainly the problems of termination merit a chapter to themselves.

CHAPTER 12

Selection by abortion

What is being aborted – The world situation – UK and US
laws – 'Right to Life' campaign – Influence on other codes –
Do-it-yourself abortion – The future

*'The simplest way of dealing with the problem is to repeal all
abortion laws. We do not have to have laws regarding abortion
at all'*

MARGARET MEAD

*'No woman wants an abortion. Either she wants a child or she
wishes to avoid a pregnancy'*

LETTER TO *The Lancet*

*'I did not vote for 100,000 abortions in the private sector, nor
for fortunes being made by the most dubious of medical people,
nor for 50,000 desperate foreign girls to flood into London'*

WILLIAM PRICE, MP

'Do Not Take Life that I Have Given. Vote "NO" on Proposal B'

A MICHIGAN POSTER DEPICTING CHRIST ON THE CROSS

'I am not here as a Catholic or as a priest but, I guess, as a grown-up
foetus,' said Dean Robert F. Drinan (of the United States) at the start of a
conference. Abortion, by any definition, is the destruction of a human
being that would otherwise have lived a life on earth. At the tenth week
of pregnancy, a common time for termination, the mother is not greatly
changed. Her average weight gain is about 1 lb., her placenta weighs
about an ounce and her uterus contains an ounce of amniotic fluid; but,
attached to that placenta and surrounded by that fluid, there is a quarter
of an ounce of foetus. Its head is three-quarters of an inch long, its trunk
about the same, while its legs are yet smaller and extremely spindly. Its
mass is not great, for a baby is 500 times more massive at birth and the
final adult is 10,000 times heavier.

Nevertheless it is not some vague protoplasmic blob, nor a portion of undifferentiated tissue. It is a human child with, if uncurtailed, only twenty-eight weeks to go before its formal entry into the world. At nine weeks it can even be recognized from the outside of its body as either a boy or a girl. Its fingerprints have already been indelibly engraved for life. It can move. Its palms are sensitive to touch. Its nails have begun to grow. Its eyelids are protruding over its eyes. Three weeks later, and without great difficulty for its mother, the foetus may still be aborted abruptly from its cosy aquatic niche in the womb; but at that time the young human has learnt how to frown. It can move its thumbs to touch its fingers. It has begun to urinate. Its vocal cords, although months away from making their first bleat, are complete. Its heart pumps a total of fifty pints a day, albeit within a body still weighing less than an ounce. It is emphatically not a blob, nor some fish-like intermediary reminiscent of our evolutionary past. It is a very small human being about to start upon the phenomenal growth that dominates the last two-thirds of a woman's normal pregnancy.

Nevertheless, for the coin's other side, its very capabilities can act disastrously in its disfavour. It will, unless stopped, become an independent offspring. It will cry and demand food. It will have rights and need a place to live. The very thought of its birth can be abhorrent, or wearisome, or merely unappealing. It can be total joy, or it can mean misery. Women bearing an undesired foetus have, in their millions, taken desperate measures to rid themselves of the incubus within them. The old remedies are frightening even to read. The possibilities of air embolism were always present. The consequences of failure, of wishing to destroy but of losing the opportunity, can be equally terrifying – the arrival of a tenth appetite when nine others cannot be satisfied, or the birth of just one dependant to a girl for whom all other prospects are thereby negated, or just the creation of one unwanted baby who then has to suffer the trauma of being unwanted throughout infancy, throughout childhood, throughout adolescence.

Nobody wants abortions, but tens of millions suffer them. No doctor, sworn by his calling to assist life, can relish the task of destroying foetuses. Inevitably the destruction has an effect upon all parties concerned, upon the mother, upon the father, and upon the doctors and nurses that become involved. It is a moral issue. It is not, as currently conducted, as simple a matter as contraception. The deliberate decisions that have to be taken today for birth control via abortion smooth the way, or so it can be argued, for yet more positive aspects of family control, for aborting conceptions discovered during pregnancy to be malformed, for aborting foetuses on the off-chance that they have been damaged by an attack of, say, German measles suffered by the mother.

Still further in the future lies the destruction of foetuses because they happen to be of the unwanted sex. And further still is a general approval of the idea that unborn children can be checked at any time throughout the thirty-eight weeks of pregnancy, checked for their quality and then, perhaps, checked from living. Should that custom become prevalent, should the idea of pregnancy rejection become acceptable whenever an unwelcome attribute is detected, will it inevitably be followed by a licence to kill the finished product at birth if it then appears unwholesome? Many a society has strangled malformed babies the moment these first saw the light of day. Even twins have been made to pay this price for their duplicity, it often being argued that two conceptions must signify two lovers. Individual decisions are now being taken at birth about spina bifida, about mongolism and other errors of metabolism. Will decisions soon be made concerning lesser errors, and will we then accept only the very fit as suitable company for our time on earth?

With abortion it can be assumed that 98 % of the expelled foetuses would have been normal had they been born. They were aborted because, in general terms, they were unwanted. Many unwanted pregnancies can lead to wanted babies, the alchemy of birth creating dramatic alteration of opinion, but the practice of abortion does not permit this change of heart. As a society we therefore both expel good foetuses and simultaneously lavish much care upon malformed disasters that have been permitted to reach full term. We warmly wish all offspring to be wanted but do nothing to kill those misshapen children that cannot be cared for by their parents. Living out their lives in institutions there is no real place for them in our world. And yet they live, while good foetuses die in their millions.

One reason put forward for the abolition of execution was the effect of such wilful murder upon the rest of us, rather than the victim. Foetuses may be minuscule when aborted, but Britain alone is throwing away more than a ton a year of these proto-humans. Such a massive act, involving so many, cannot be irrelevant in its effect upon our ethics, our judgements, our future policies.

So what is the present state of abortion in the world? Legally or illegally it is practised everywhere. More than half of the planet's population lives in countries permitting abortion for social as well as medical reasons. In effect this was the case everywhere until the start of the nineteenth century. Britain had no laws against termination of pregnancy before the 'quickening', the foetal movement first occurring during weeks 18–20, until 1803. In the United States prohibition of such abortions did not arise for another half-century. However, the early attempts at evicting the foetus cannot really be compared with today's capabilities. To use slippery elm bark, to fall deliberately downstairs, to drink gin with iron filings, or potassium permanganate, or Widow Welch's recipe, is not the same as to walk into some officially approved clinic, without fear of either failure, stigma or death. In fact it was death, the common result of careless jabbing with knitting needles, or the remorseless infusion of violent remedies, that provided the spur for all the anti-abortion laws of the last century. It was protection of the mother rather than curtailment of her freedom which caused the restrictions that so much of the world, thanks to safer abortions, is now trying to unravel.

The Soviet Union, freed from virtually all restraint by the revolution of 1917, permitted abortions for almost any reason three years later. The Communist countries of eastern Europe followed suit in the 1950s. Japan, under her eugenic protection law of 1948, attacked her population problem as decisively as everything else; abortions rose to a peak of 1,170,000 in 1955 and have since declined to two-thirds of that number. In Britain a few years ago the official figures were both totally misleading and of quite a different order. In 1955, for example, there were hardly any legal abortions and only 66 recorded as resulting from 'criminal interference'. In 1968 the laws were liberalized under an Act permitting more of reality to come to the surface. In the first four years after the new Act there were, respectively, 23,000, 54,000, 86,000 and 127,000 legal abortions, but these figures were being distorted by Britain's neighbours. In 1971 over 31,000, and in 1972 over 51,000, of the abortions carried out in Britain were on women 'normally resident elsewhere'. Most of these visitors, helpfully swelling the tourist figures for those years, and representing a total of 820 planeloads at 100 seats a plane, were from Europe. Any country temporarily ahead and more liberal than its neighbours will experience such an influx. 'Fly her to Tokyo,' said the Japanese to the Americans, 'and leave the rest to us.'

That particular trade with the Orient must now be dramatically reduced. Abortion reform in the United States started later than in Britain but then went faster and further. In 1970 only four states had totally repealed their nineteenth-century laws, but state legislation can be challenged federally if it is felt to impede an individual's rights under the constitution. The laws of both Texas and Georgia were held (on 22 January 1973) to be unconstitutional by the Supreme Court because they imposed restrictions upon the rights of the pregnant woman.

The Court then laid down basic principles. Up to the end of the third month, or thereabouts, the woman's doctor – and the woman – could decide whether or not to abort. Immediately thereafter, in time and the extent of the pregnancy, 'the State, in promoting its interest in the health of the mother, may, if it chooses, regulate the abortion procedure in ways that are reasonably related to maternal health'. Thereafter again, and when the foetus is viable – after the sixth month – the State may proscribe abortion, save where it is necessary for the health or life of the mother. In other words, riding more loosely over the precise language of the Court, a woman now has the same right to an abortion during the first six months of her pregnancy as she has to any other surgery. For the last three months the foetus has rights, although less so than when it is actually born. The ruling is so bold and uncompromising, as *Time* magazine put it, that even the four states with the most liberal laws (New York, Washington, Hawaii and Alaska) will probably have to make some changes.

Available facilities are also due for some alteration. Based on New York's experience over the past three years, and according to Christopher Tietze, of the Population Council, provision will have to be made to provide abortions for 1,600,000 women a year. It is now incumbent, he said, to make abortions available to all at a fair price. Presumably that will be nearer the current outpatient clinic cost of $100 rather than the $400–$600 charged by some hospitals. In any case the abortion business is certain to be a major industry because 1½ million women a year at, say, $200 a time means an annual gross of $300,000,000. There is no point in any additional freedom, or rather any new statement on the right of privacy implied by the constitution, if it costs too much.

One wonders whether the Founding Fathers would wholly concur with today's latest interpretation of their hallowed words, but the Supreme Court's thunderbolt has undoubtedly arrived. The men to fire it – by a majority of seven to two – were a group of people previously thought of as largely conservative. Nevertheless, many others feel the move to be entirely reactionary. 'It is a step backward', said a bishop of the United Methodist Church, 'if the new ruling leads to promiscuity and to taking the creation of life lightly.'

What will the new ruling lead to? How will the liberalization of abortion affect fornication, a regard for life, or ethical judgement? Essentially the laws have not created something entirely new; they have merely brought a concealed situation into the open. France, for example, is in a confused and concealed state, ripe with deceit and reaction. In 1973 Michelle Chevalier was fined £40 (the maximum penalty for the offence being £6,000 or ten years in prison) because she had obtained an abortion for her raped 16-year-old daughter. Even the daughter was tried, although acquitted, in a country where estimates on abortion often put the total as high as the birth rate. Such a *cause célèbre*, which had Madame Chevalier's counsel asserting that the verdict would 'toll the knell for the infamous Law of 1810', makes the French situation similar to that of Britain in 1938 when Aleck Bourne, the obstetrician, performed an abortion on a raped 14-year-old. At least, after his lengthy trial, he was acquitted, whereas the Chevalier abortionist was given a year's suspended gaol sentence.

In 1972 Belgium gaoled the respected gynaecologist Willy Peers for performing 300 abortions in his Namur clinic. In Latin America, where Catholicism and the pronouncements from Rome certainly discourage new laws on abortion, there is similar double dealing, with millions of abortions annually, but with few people brought to book for their law-breaking. Brazil alone reckons one million abortions a year, and Chile believes that one-third of all her pregnancies are prematurely terminated. Court cases are rare, fines are generally light, and the law is not so much an ass as a machine incapable of dealing with the hundreds of thousands of frightened, desperate and sick women, who are all the more so if the foetus has been replaced by infection. The World Health Organization estimates that the world abortion rate is thirty million or more a year, with practically all of it illegal. It may be macabre to think of this total by weight, but millions of anything are always difficult to visualize. With each aborted placenta weighing an ounce and each foetus a quarter of that amount the annual and disposed quantity estimated by the WHO adds up to 837 tons of placenta and 208 tons of foetus.

India achieved a new abortion law in April 1972, and within a year 15,000 Indian women had had legal abortions as against the four million believed to have occurred illegally in the same period. The president of the Lebanese Order of Physicians has called the abortion rate in his country 'an epidemic', but confronted by Christianity, Mohammedanism and mere tradition the outbreak is likely to continue. Or perhaps, on this most private and important issue, the climate of opinion will suddenly

195

shift. In the United States, now aborting liberally, people were being gaoled merely for sending contraceptive literature through the post a few years ago. A British Home Secretary, admittedly not the best of his kind, was stating in 1964 that 'the time was not yet ripe' for amending the abortion law. It was very ripe by 1967 when Parliament drafted and passed the new Bill.

Is it likely therefore to become overripe? Will public reaction swing against this form of contraception, particularly when premature babies can survive much earlier than today? The abortion issue is not yet dead and buried, despite the welcome given to new laws, despite the annual relief felt by thirty million women (and presumably thirty million men) as their unwanted problem is removed. In Britain today there is more activity denigrating the liberal abortion regime than at any time since the laws were reformed in 1968. In May 1972 some 40,000 supporters of the Society for the Protection of Unborn Children met at the Liverpool Pier Head. In the following spring the same society organized a yet bigger rally in Manchester, the kind of march where the leaders are on their way back home before the end of the procession has even begun. All participants wore white cardboard flowers, each representing an unborn child killed under the Act.

'Over 450 lives are being aborted every day, and there can be no possible excuse for carnage on this scale,' said a speaker. Malcolm Muggeridge, the radical reactionary (or vice versa), denounced those who saw 'our generation as just a productive process, to be fostered or abated like any other poultry, or pigs, or mangel wurzels, in accordance with our material circumstances'. With less rustic phraseology, but making the same kind of point, New York's Archbishop Terence Cardinal Cooke called the liberal laws 'a tragic utilitarian judgement' and added that 'judicial decisions are not necessarily sound moral decisions'. Right to Life's chairman in Georgia, Joe Bowman, realizing the foetus's poor lawful lot these days, was reminded of the 1857 Dred Scott decision which said that the black man, although he had a beating heart and a functioning brain, was not a legal person.

The US Supreme Court's 1973 decision is not immutable. (Right to Life organized a major national demonstration against the decision on its first anniversary, culminating in a march on Washington.) If Catholics and other enemies of abortion press sufficiently strongly for a constitutional amendment the various States could overturn the revolutionary ruling. As David Laulicht of Reuters said, after listing current global opinion, the ultimate meaning and outcome of that Court decision are still as uncertain as in all other revolutions. Incidentally Russia, having – in 1920 – been the first to liberalize abortion, switched in 1936 to make it extremely difficult, and then switched again in 1955 to make it easy once more.

The back-swing may or may not gain sufficient momentum to amend the new laws, but it undoubtedly exists. William Price wrote to the *Guardian* in April 1973: 'I am one of those Members of Parliament who, on balance, supported the Abortion Bill and no single act during seven years in the House of Commons has caused me more personal anguish. Today I cannot say whether I was right or wrong ... I ask your readers to believe that some of us in the Commons are having serious trouble with our consciences.' Even the Japanese, with no more land than before and with more people, are talking of hardening their abortion laws by removing the economic clause.

Those promoting the back-swing may find it impossible to de-liberalize the modern law and may therefore exercise their consciences, their regrets and their religious opinions upon the next piece of similar legislation to arise. Besides, as Governor Nelson Rockefeller said, 'repeal would not end abortions; it would only end abortions under safe and supervised medical conditions'. It would be easier to divert one's principles, to attack any suggestion that, for example, foetal sex might be controlled, that amniocentesis should be pursued for the ultimate rejection of undesirable traits, or that murder of grossly malformed babies born at term should be legalized. Other changes now waiting in the wings both for the necessary medical advance and for legislative sympathy may be rebuffed with extreme vigour when the time comes for their public airing, with current liberal abortion being the sole cause of these determined rebuttals.

Something of the antagonism smouldering beneath the issue of pregnancy termination came to light in a *Daily Express* (of London) story: 'Abortion Trade in Unborn Babies' ran the headline with, beneath it, 'The Poor Creatures who are denied Life under the Abortion Act are then to be denied Death'. Norman St John-Stevas, MP, always opposed to the Act and rarely short of words, was quoted: 'The practice is the most horrible that has ever taken place in Britain. It is almost unbelievable that this could be happening in a civilized country such as our own. The thought of what has been going on makes me feel physically sick.' The alleged practice was the selling of live foetuses for medical experiments. An immediate result of the story was that the relevant minister, Richard Crossman, promptly forbade private abortion clinics from supplying foetuses to research centres but permitted the state hospitals to do so. This distinction was presumably guided by basic distrust of anybody doing anything for money, however much all British abortion clinics have both to be officially approved and to make ends meet financially. The *Guardian* ('Clinic Sells Abortion Babies for Research') and *The Times* of London ('Live Foetuses Sold For Research') implied by their choice of words that money changing hands was a disturbing feature of anything else that might or might not be happening.

A parallel medical story has never encountered this kind of furore. Whether their relations know it or not, and whether the victims were ever informed earlier or not, the pituitary glands of more than one-tenth of the people dying in England and Wales are removed fairly soon after death. This particular tissue is essential for the growth management of all those people incapable of making sufficient pituitary hormone for themselves. Without enough of it they will never grow correctly; with enough, however it comes to them – and the substance cannot yet be created artificially – they have a better chance of normal stature. To make use of any tissue from any cadaver is plainly desirable if the dead can therefore aid the living, and there has been no difficulty over the pituitary extraction. Money may even be involved if private clinics are helping to fulfil the need. Yet there was controversy over the supply of foetal material, however non-viable on its own, if only because the story helped to prey upon the dislike, and uncertainty and guilt-ridden feelings, that are undeniably linked with abortion, with this most brutal form of human contraception.

The *Express* storm died away and within two years a doctor was even able to recommend that foetuses about to be aborted should be given drugs whose effects, teratogenic or otherwise, could be examined later on. The suggestion was from William McBride of Sydney, the Australian who had been the first in his country to recognize the link between thalidomide and limbless babies. As of now aborted foetuses of women who have had German measles during pregnancy are subsequently examined. Therefore, as the taking of drugs is a good deal commoner and plagued by more unknowns than the catching of this lesser form of measles, it is reasonable – or so McBride argued – to try out drugs on doomed foetuses. 'Should not the condemned be used to the benefit of future generations of children?' he asked, however much the logic of medicine does not necessarily coincide with prevailing ethical sensitivity among the public.

The importance of abortion will continue, both as an opinion-maker and as a doubt-raiser, so long as it exists in its present form. As one example, take its association with the population problem. It is agreed that this is a problem. It is agreed that countries need population policies, that practically everywhere must hone down its current birth rate. It is also agreed that compulsion is a poor method of family limitation, and coercion via propaganda or financial inducement is preferable. And finally it is generally realized that contraceptive campaigns provide a good investment. Money spent now means a saving in state expenditure later on, when there is one less mouth to feed, to care for, to educate, to house.

Therefore what about the proposal, put forward in good faith from various quarters, that women should be paid for having abortions? In the United States and various other non-welfare communities they have to pay. In welfare Britain and communist societies the treatment is free. Suppose, instead of a bill of £150 a time (the old high-quality British price for illegal termination), that all women electing to have any future abortion were paid £150 a time on leaving the clinic. It would still benefit the state. The prevention of, say, 100,000 babies would cost £15 million, and that kind of money would have been quickly absorbed in providing for the needs of 100,000 children. At 800 children to a school, quite apart from a reduction in all other facilities, the requirement for 125 fewer schools would be a boon nation-wide. Money saved would be more than £15 million on this single account.

The number of women likely to exploit the system, by having three or so abortions a year for the sake of £450, would surely be few, but the arithmetic would still be satisfactory from the state's point of view. So what about world opinion? People are, in general, welcoming more liberal systems of abortion. Many people are in favour of such liberality without payment. And yet, or so surveys say, we would almost all be totally against remunerative abortion. Perhaps change on this score will come sooner than we think. Perhaps, increasingly aware of the tide of over-population, we will accept any measures likely to stem the flood. Perhaps abortion will lose every last ounce of its unfortunate taint and become a wholly virtuous act, a self-sacrifice by the woman, a commendable deed for the good of the community.

However much some may argue that all forms of contraception are similar, whether they prevent the union of sperm and egg (sheaths, caps, spermicidal ointments), or disrupt each menstrual cycle (the pill), or prevent implantation (coils, loops), or eject the foetus at a later date (abortion), it is the majority opinion that termination of pregnancy when the foetus is well formed involves greater ethical problems than other forms of birth prevention. 'Abortion Kills' say the rally placards – and so it does. Words like kill and murder become, for most people, increasingly valid with the increasing age of the conceptus, but pregnant women first have to realize their pregnancy. They then must decide what to do and, to a greater or lesser extent, they are then kept waiting by the system – appointments, examination, clinic vacancies. In liberal New York, and in 1970–1, only 73 % of all abortions were carried out even in the first three months of pregnancy. Abortion will surely be less repulsive if it happens earlier; indeed, the ratio was up to 82.2% two years later.

The fear of death for the mother is one factor helping to reduce this lateness. In New York there were seven deaths among the 321,500 women who were aborted at less than twelve weeks (between 1 July, 1970, and 30 June, 1972), a proportion of 2.2 deaths per 100,000 cases. During the same two-year period there were fourteen deaths among the 80,500 women who had abortions at thirteen weeks or more, a proportion of 17.4 per 100,000. The overall death rate is therefore 5.2 per 100,000, and it is certain this can be reduced if women are more prompt in requesting the operation and if the operators are swifter about their business.

Britain's death rate from abortion has followed a similar pattern. The uncertainty, not to say the downright difficulty, linked with abortions before the Act of 1967 (which became law in 1968) must have brought women to their abortionists at a later stage than now. Deaths were thought to have been about thirty a year in the early 1960s, about fifteen in 1969, and eleven and six in the next two years. Promptness has become so much easier and the combined death rate for abortions of all kinds in Britain has slumped from always above six per 100,000 in the years before the new law to four per 100,000 in 1970 and three per 100,000 in 1971, as against New York's 5.2.

To be early is better. The operation is simpler, the trauma less. The destruction is more like contraception, less like murder. In fact, extreme earliness may well become the pattern for the future rather than the ten or so weeks of today. For example, there is menstrual extraction (called endometrial aspiration in America). Harvey Karman, the Los Angeles psychologist, was the man responsible for the famous cannula technique, whereby this instrument is inserted into the cervix and the uterine contents are sucked out, with a vacuum pump creating the drop in pressure. The new method is to use a similar device, so simple that anyone can handle it successfully by following the instructions, the moment a menstrual period appears to be overdue. It is cheap, it is easy, but the extractor must be sterile. For this reason, despite its simplicity, said Malcolm Potts, the 'equipment should not be pushed out on to the general market like Coca-Cola'.

As always there are potential drawbacks. What are the effects upon the womb of such a suction, of the events of a normal five-day period being compacted into a mere five minutes? Women do not become pregnant as frequently as they think, for periods can be disturbingly erratic in their behaviour and the women can be disturbingly faulty in their calendar arithmetic. Therefore will they reach for their cannula at the first doubt? Will they forget the pill and simply induce their menstruation? And will the technique, by expelling a three-week embryo one-tenth of an inch long, be considered ethically more satisfactory than aborting a three-month foetus two and a half inches long? Might it even mean that the word abortion will slip into the past and menstrual extraction take its place?

As an extreme by-product the device may reduce the intense problem of conscientious objection to abortion. In a letter to the *British Medical Journal* (in October 1972) a doctor reported how 'he had been advised to cut his losses and emigrate'. The head of a teaching hospital had told him there was now no place in Britain for any gynaecologist to practise within the National Health Service who objected to abortion on demand. Young Roman Catholic doctors are therefore being warned to avoid the speciality of gynaecology for this reason. If menstrual extraction is thought to be less objectionable, at least this dilemma might be reduced. After all, most Catholics in Britain, when polled to this effect, say they not only favour but practise contraception. They do not appear to resent the coil, which may well be evicting an unimplanted and six-day embryo, and might therefore see little difference in the expulsion of a developing form only fifteen days more mature.

As a final point it is interesting to reflect that any apparent improvement in the human lot is sure to be pounced upon by someone as a preliminary to dire consequences. Abortion is no exception. Margaret and Arthur Wynn, having surveyed medical publications from twelve countries with experience of legal abortion, reported in 1972 that: 'It would be wise for young women and their parents and future husbands to assume that induced abortion is neither safe nor simple, that it frequently has long-term consequences, may affect subsequent children and makes young single women less eligible for marriage.' They quoted research studies in which 2–5 % of women having abortions became sterile as a result, while those who did achieve subsequent pregnancies had more miscarriages and more premature births than normal. The babies themselves had a higher than normal risk of mental or physical handicap. Having read the Wynn report others were then free to debate the evidence and the conclusions, and the argument still continues with a certain erosion of the original statement.

Nevertheless, it is intriguing that medical follow-up news tends to be unfavourable. Of course smoking, so pleasurable to so many, would be found to be harmful. Of course rich and exuberant foods are linked with every manner of malfunction. Pondering further, it is intriguing to wonder whether any welcome human antic has ever had association with an unexpected and *beneficial* effect. Might over-eating, for example, unaccountably help to cure ulcers? Might sunbathing, apart from being carcinogenic and generally damaging to the skin, not do more good than give us vitamin D of which we normally have enough and can suffer from having too much? It is just a thought and not irrelevant for future reproductive changes.

Nevertheless, whatever happens in abortion will affect society's attitude to events in other spheres of biological manipulation. It will harden us to change, or render greater enthusiasm, much as a crack can bring down a wall. It does not have the simplicity of contraception. It is all very well to state, as so many of us (and the Supreme Court) are doing, that life begins either at one moment or at another but, as C. B. Goodhart said: 'We need not resolve the difficult question of when life begins. When those trained in medicine, philosophy and theology are unable to arrive at any consensus, the judiciary at this point in the development of man's knowledge is not in a position to speculate as to the answer.' But we will speculate, all of us. We will increasingly adjust our reproduction, given the medical means and the legislative power to do so. Abortion, in some sense, showed us the way, but which way is quite another matter. Perhaps we will next wish to choose the sex of our offspring, perhaps not, but this topic and this possibility undeniably form the subject of the following chapter.

CHAPTER 13

Sex ratio

Current numbers – Sex determination – Sex control –
Masculine preference – Polyandry – Sex-linked disease –
Future implications

'When a girl is born, the walls are crying'

TALMUD

*'When an Arab hears that a daughter has been born to him his
face becomes saddened'*

BOOK OF ISLAM

'It takes a man to make a girl'

COCKNEY SAYING

As with so much of biology there is enigma. The sexes are fundamentally
different, right down to the quantity of red blood cells per cc, verbal
ability, cerebral structure, and whatever aspect, or so it would seem, is
investigated. (Read Corinne Hutt on *Males and Females* for considerable
enlightenment in this area.) But the sexes, it can also be argued, and with
equal facility, are basically identical. Each of us produces male hormones
and female hormones, only in differing amounts. Each side can, via a
natural or injected shift in the hormonal balance, produce the secondary
sexual characteristics of the other. And, yet more remarkable, adult men
and women can change sides, with surgery providing the necessary means
for this greatest volte-face of them all. We are, all of us, alike and different,
immutably fixed from the moment of conception and forever susceptible
to change.

Environment can also manipulate much of genetic inheritance. It can
certainly amend our sexual endowment, transforming expected hetero-
sexuality into like wanting like, but basically the inheritance is totally
genetic. It is the famous story of the X and the Y, first encountered in
humbler creatures – such as mealworms – at the start of this century, and
then belatedly found to follow suit in mankind itself. Only in 1956 was
the human chromosomal count established for all time as forty-six, with

twenty-two identical pairs and those two extremely dissimilar oddities, the sex chromosomes, the Y and the X, the male and female kinds.

The male's inheritance of one X and one Y, and the female's inheritance of two X's, at once underlines both the fundamental difference and the similarity. A male's X is identical to both of those of the female, and each male must have this female component as well as his Y to be normal. The female cannot rely for normal development upon either the presence of one X or the absence of the Y; there have to be two X's in her genetic make-up for normality. It is for this XX reason that the female does not control the sex of her offspring: she can only pass on an X. The male, by being XY, produces two kinds of sperm: the X sperm that will create females and the Y sperm that will make further males.

Normally XXs and XYs are created in roughly similar numbers. No one knows yet how this is achieved, but there is a major wish to amend its inevitability. Like the promise of controlled thermonuclear power, or a vaccine for the common cold, we have all been informed for a good many years that the age-old ratio of boys to girls will soon be adjustable. Like a carrot it dances before us, and sheer persistence in the laboratories will surely cause us to catch it, to discover a means at last for altering the numerical equality between the sexes.

First, then, to the existing ratio and further strangeness. For reasons no one has yet begun to comprehend the allegedly stronger sex is weaker all along the line. Many more males are miscarried both early and late, more males are stillborn, and more males die between birth and maturity. As compensation for this weakness it has been suggested that more males are conceived, perhaps 150 for each 100 females, but there is no evidence for this. Anyway, by the time of birth the proportion is about 105 to 100 in western societies and either higher or lower in other parts of the world, ranging from 101/100 to 113/100 but without any discernible logic. It might be thought that poor conditions would harm a greater number of the more susceptible sex, reducing the ratio towards parity, but Greece, for example, a community impoverished by European standards, has one of the highest ratios of all.

The production of one of each sex, particularly in our kind of society, seems entirely reasonable. The disproportion of males to females at birth seems yet more reasonable when we learn that, owing to a continuing masculine weakness throughout childhood, there is sexual equality in numbers at the age when it really matters. However, the simplicity is misleading. Why parity? Why, in those early cave days, could there not have been a preponderance of women, breeding continually to compensate for considerable losses, and readily fertilized by a mere handful of men? Or, conversely, why was there not an abundance of men, suffering death frequently at the teeth and claws of every hunted thing? And how strange that our inheritance, fashioned in ages and societies so very different from our own, should be so eminently suitable here and now.

The general evolutionary picture is also intriguing. How does natural selection, which must be the arbiter of the sex ratio, actually work? It is populations rather than individuals that experience selection pressure, but it is through the individuals – producing offspring of different ratios – that it has to work. Suppose that disease or other circumstance suddenly begins to kill off many young boys. Suppose further that the society adjusts for this lack by suitable polygamy. By what manner of feedback would a greater number of boys be born in the future? There must have been occasions when women were at a premium, following maternal deaths and selective disease – and there must have been times when males were more vulnerable – warfare, puberty rituals, greater susceptibility to some other hazard; yet, in our extremely various societies, ancient and modern, the sex ratio at birth is always slightly in favour of more males. So too, if a sweeping generalization is made throughout the animal kingdom, with most other vertebrates. There is usually parity but, if there is dissimilarity, it is almost always in favour of the male.

Such a well-rooted tradition has not, of course, been partnered by a universal human wish to leave things as they are. The lure of being able to control an offspring's sex has an ancient history. Right testicles for boys, left for girls, said the Greeks. Therefore accept the discomfort and tie accordingly. Right ovary (or right uterus for the days when they believed in two uteruses) for boys, left for girls, said others. Therefore contort appropriately during coition. Hang your garments on the bed's right side for boys, and left side for girls, said and still says much of rural America. Therefore think twice before stripping and hopping into bed. It is interesting that the universal prejudice about the right-hand side – as opposed to sinister, widdershins, leftist, morganatic, non-dextrous – is unfailingly used in favour of masculinity.

In recent times relatively more males have been encountered among the offspring of mixed race marriages, AB blood-group mothers, radiated fathers (in Japan in 1946), very old mothers, very young parents, wartime or immediately post-war liaisons, scientific staff and engineers (as against non-scientific staff in the same organizations), and pilots who fly the slower as against the high-performance types of aircraft. All very odd. So too, according to *Vital Statistics of the United States,* that the sex ratio shifts from 104.4/100 for October–December births up to 105.8/100 by midsummer. Intervening months are of intervening ratios, and the kind of change applies to both white and non-white even though the Negro ratio normally averages nearer 102/100.

The merest glimpse of contradictory confusion is enough to spark many of us into searching for some elementary explanation. Landrum B. Shettles and David Rorvik (in *Your Baby's Sex*) produced a general statement, conveniently covering a good many of the requirements, that male-bearing sperms do less well in the more acidic female environment of the vagina and cervix just before ovulation and rather better when it becomes more alkaline a day or so later. Therefore abstain from intercourse until later in the month if boys are required. Do, as good Jews do, and wait for one week after menstruation, and then create more boys – as Jews do. Or just wait a bit and up goes the proportion of male-bearing sperm, a fact which accords less well with the higher male proportion born to young parents. Artificial insemination, which is much less casual than mere intercourse in aiming for the time of ovulation, allegedly creates more boys than girls; a British clinic gave the proportion as 3 to 2 or, to use the familiar style, an unparalleled ratio of 150/100. One feels that AID, without in any way prejudicing its basic aim of making women pregnant, could settle this argument once and for all of when to conceive. So far it would seem to be unproven, or would not more have been heard of it?

If a blue pill for boys, or a pink pill for girls, is eventually achieved it will plainly be more satisfactory as a sex determiner than the timing of intercourse or the seeking out of suitable mates, the fast fliers, engineers and so on. But will it, as many have argued, be a suitably ethical and immediately effective answer to the population explosion? At first sight it appears extraordinarily attractive. John Postgate, British microbiologist, has said that 'a pill which favours fertilization of the human ovum by a sperm bearing a Y chromosome could well be mankind's only real hope of becoming once more a stable part of the Earth's ecology'.

In most underdeveloped nations the male child is the desired child. In many primitive societies the girl was often killed along with malformed products and twins: the Eskimo was said to place her under the ice. In Africa, Asia and Latin America today the wish for a boy is still strong and much like an obsession. Even in Europe and North America the polls suggest that the most welcome sex for a first-born is male. In short, a male-only pill, according to the prevailing notion, would not have to be linked to campaigns, propaganda and transistor radios, as with family planning, but would fetch the highest prices in every bazaar throughout the world. Think of some Mohammedan wailing at the news of his third daughter, at his lack of a man-child; in fact, think of any Mohammedan. Or think of a Hindu, for only a son should take part in certain cremation rituals of his parents. And then think of their faces, and their haste, should they hear of 'Androcert', the male guarantee, on sale now just down the road and at a price within reach of their credit capability.

The important fact for population growth is, of course, that babies can only be born to women. Therefore, if the average family in, say, Egypt becomes three boys and one girl, the total number of offspring will be four but the real number, the breeding number, can be taken as two, or one girl and one boy. In under-privileged countries, still rich with sickness, the population replacement level is currently about three children per married couple. Therefore, three boys and one girl per family would actually be below replacement level and, in time, the population would shrink. To achieve 1.5 girls per family, still assuming a masculine preference of 4:1, and to acquire thereby the necessary number of women to produce a stable situation, the average family size would have to leap to 7.5 children, or 6 boys and 1.5 girls.

Even in Europe, where the masculine preference is less strong, the effect could also be dramatic. Currently, the average family of 2.5 is divided equally (well, almost) between boys and girls. Should such male preference as exists shift this division to 1.1 girls and 1.4 boys the result would be a Europe no longer afflicted, once levelling off has occurred, with a growing population. In short, Androcert would have an influence to eclipse even that of contraception.

One British survey (by John Peel of York University) asked young couples to state the sex preferences for their future offspring. The answers from the sample of 350 couples living in Hull need to be given in full. Hoped-for sex distribution and family size was:

1 boy 1 girl (preferred by 160)   3 boys (by 5)
2 boys 1 girl (by 54)   4 boys (by 5)
2 boys 2 girls (by 44)   1 boy (by 5)
2 girls 1 boy (by 24)   1 girl (by 3)
2 boys (by 11)   none (by 6)
2 girls (by 7)   other combinations (by 14)
3 boys 1 girl (by 7)

As the mean intended family was 2.6 children, and as the normal sex ratio of 106/100 (in the area) would normally lead to 477 boys and 435 girls – assuming everyone achieved their intended family size, the shortfall of boys would only be fourteen. 'Thus an increase,' says the report, 'of only 2.9 % in the ratio of male to female births would be sufficient to give all these parents the precise sex distribution they require.' But that is for the surveyed population as a whole; few couples are likely to achieve their hoped-for ratio – without the aid of pills. It is also notable that, of the eleven specified kinds of families, nine included boys and only six included girls. Yet more fascinating would be the result of similar surveys from the areas around, say, the Nile, the Ganges or the Tigris.

Malcolm Potts rejects the idea of a sex-determining pill. Firstly, it is difficult, he says. Commercial agriculture has tried to control sex for years but has failed. Secondly, the necessary drug testing would put 10–15 years between development and use. Thirdly, the medical establishment, already 'providing the greatest single barrier' to oral contraception by insisting on prescription almost everywhere, would be even more awkward over 'something as controversial as a sex selective pill'. He feels too that parents would resort to such a pill only after a daughter, or two, or three, had been born, and the sexual imbalance hoped for by John Postgate and many others would not arise to any desired extent.

Nevertheless it does seem inevitable that a sex-determining pill will arise. It might have unwelcome side-effects. It might be expensive. It might, officially, be on prescription; but suppose for a moment that the souks of the world, by disregarding the dangerous drug policies in more inhibited areas, succeeded in distributing the pill. Suppose then that daughters become rare, with three boys and one girl per family a commonplace. What then, when the parental jubilations favouring masculinity are over, of the next generation? Will girls vanish from sight, losing such emancipation as they have acquired, and be offered at high prices as the only begetters of further males? Will polyandry be welcomed or tolerated anywhere, let alone in societies traditionally geared to polygamy? Postgate is not greatly concerned, and adds: 'However unpleasant a rapidly declining, largely male society might seem, altruism, kindness and co-operation would still stand a chance; we know they do not have a bat's chance in hell if the world goes on as at present.'

Presumably there would be more homosexuality, more substitutes for sex and a transitional situation abhorrent to many, but perhaps the male preponderance would only be temporary. Something equivalent to the commercial law of supply and demand might arise, quelling the machismo, boosting its converse, and causing the bazaars to think pink, not blue, for a time. Looking even further ahead, everything might eventually settle down as 105/100, the magic formula of today. Even if such normality did return, and if the intervening confusion was hell for many, the experiment would certainly have jerked the current population graphs from their remorseless and exponential trend. To stop the world for a moment, to create a breathing space, however socially confusing, might provide the kind of numerical respite for which we are searching.

Some sociologists cannot welcome the social disturbance, however welcome the population respite. An American report (by A. Etzioni in *Science*) bases its concern on the assumption that 7% more boys would be demanded, as against Hull's 2.9%. It predicts a 'socially disastrous' time caused by an annual surplus of 350,000 boys in the United States. It therefore suggests that biological research on sex determination should be controlled and 'if necessary prevented'. Science has never been skilful in stopping intriguing lines of research and few biologists could either tolerate or imagine how such a block might be imposed that would forbid the acquisition of further knowledge concerning X and Y, the very kernel of sexual reproduction.

So when, as with the cold vaccine or the fusion of deuterium, will a sex determinant ever come about? At present the main effort is to detect the sex of a foetus as early as possible. This form of amniocentesis is aimed at the early discovery of potential victims of sex-linked disease – haemo-philia, dystrophy – so that they can be aborted in good time and with the minimum of distress. In 1969 a Swedish team discovered, after staining chromosomes and then shining long-wavelength ultra-violet light on them, that they fluoresced differently in different places. In 1970 it was reported, again from Sweden, that the ends of the human Y chromosome shone more brightly – how convenient life can be at times – than any other part of all the other chromosomes.

The detection of foetal sex, when warranted, is now possible with a high rate of success. No one in authority has yet seriously advocated the procedure for mere parental interest, for knowing six months in advance whether the growing lump is the longed-for boy or girl or not. One suspects that such investigation will occur, but it is also probable that early sex discovery, remembering the size of the uterus and the small initial quantity of amniotic fluid, will always be difficult, fairly expensive and unlikely to be commonplace.

Claims have been made since 1961 for distinguishing sperms in bulk. Further claims have been made for separating them, but no techniques are yet generally accepted. Separation will be possible, some day, if only because of that disparity between the two sex chromosomes. There is no such detectable and chromosomal disparity between the races, or between the big and the small, or the wise and the stupid, but there is between male and female. The chromosomes look different, they must be different in weight, and it has been argued that they cause a different shape to the sperm bearing them. It will not be long before these dissimilarities are sufficient for either kind of sperm to be promoted unilaterally and for an offspring's sex to be determined in advance. At least it will be better than constricting a testicle, contorting righteously, or as Pliny described, suffering a north wind to ensure the fusion, if it is your wish, of her 22X with your very own 22Y; in short to give rise to a boy.

CHAPTER 14

Eugenics past

Harry Laughlin's MD – Name-changing – Galton and the
nineteenth century – Mendelian simplicity – 1900–14 –
America's immigration laws – Compulsory sterilization –
Hitler and Himmler – RUSHA – Schultz and the SS –
Lebensborn e.V. – The Ahnenpass

'*I prefer not to quote the German law on the subject because it is
inevitable that to do so would give rise to a certain amount of
prejudice*'

J. B. S. HALDANE (in 1938)

In 1936, three years after National Socialism had come to power in
Germany, and when the racist doctrine of the Third Reich was being
made conspicuous by yet another assault on Jewry, an American was
invited to Heidelberg University to receive an honorary MD degree. This
was not just one more scientist receiving a foreign accolade in the country
that had been the greatest scientific community of them all. It was Harry H.
Laughlin, top American eugenicist, going to an establishment that
had suddenly switched its ancient political sentiments. Heidelberg had
been traditionally liberal but, when the country had moved to the right,
this 550-year-old university had been most sprightly in following suit.
The honorary MD was therefore endorsement that racist Germany, or so
it appeared, was wholly in accord with eugenic America.

The fact that the United States was already back-tracking from much
of its extremism, and the further fact that Germany carried onward to
excesses that will never be forgotten, has meant that Germany's racial and
eugenic actions now overshadow those of other nations. Nevertheless, in
the early 1930s many countries had policies which, as seen from today,
would be readily remembered but for the German example. In a sense it
was vigorous Nazi leadership which proved to many eugenist leaders
elsewhere that the path on which they also had been taking strides was not

the right one, but the steps taken before this revelation had been extraordinarily similar.

In America, and in 1931, twenty-seven states had compulsory sterilization measures on their statute books, with potential victims including 'sexual perverts', 'drug fiends', 'diseased and degenerate persons' and 'drunkards'. California had carried out almost 10,000 such sterilizations by 1935. The German Hereditary Health Law was in fact introduced by Hitler's government (on 14 July 1933) but could well have been initiated without the help of the Nazis because the idea had been actively promoted for several decades. However, the Nazi enthusiasm for preventing defective persons from contributing further to the Third Reich eclipsed the similar projects in other countries. Before July 1934 over 56,000 had been sterilized in Germany, and five times that number eventually suffered the operation. Later, and following a decree of 1939, the refinement of sterilization became the execution of euthanasia. Fifty thousand were killed in the first two years following that decree, and then the millions of the war's later years. The earlier sterilization procedure is therefore swamped by all this slaughter, but it should not be forgotten. Neither should the fact that, apart from America, which had begun by sterilizing certain state institution inmates – in Indiana – in 1907, Finland, Sweden, Norway, Iceland and a couple of Canadian provinces also had legislation to prevent various kinds of undesirable citizen from possible procreation.

At the risk of creating an exceptionally indigestible passage it may be helpful to give an exact idea of the kind of undesirability that the legislators had in mind when they drew up their laws. The rest of this paragraph, written by the same Laughlin who was so honoured at Heidelberg, is worth struggling through as its definition of a 'socially inadequate person' is surely revealing.

'A potential parent of socially inadequate offspring is a person who, regardless of his or her own physical, physiological or psychological personality, and of the nature of the germ-plasm of such person's co-parent, is a potential parent at least one-fourth of whose possible offspring, because of the certain inheritance from the said parent of one or more inferior or degenerate physical, physiological or psychological qualities would, on the average, according to the demonstrated laws of heredity, most probably function as socially inadequate persons; or at least one-half of whose possible offspring would receive from the said parent, and would carry in the germ-plasm but would not necessarily show in the personality, the genes or genes-complex for one or more inferior or degenerate physical, physiological or psychological qualities, the appearance of which quality or qualities in the personality would cause the possessor thereof to function as a socially inadequate person under the normal environment of the state.'

As J. B. S. Haldane said (in *Heredity and Politics*): 'Now you see that goes rather far!' Even the Nazi sterilization laws did not go so far.

Kenneth M. Ludmerer has written an excellent account in *Genetics and American Society* of the way in which genetic thinking, notably in the early years of this century, caused laws to be passed which are, according to current ethics, extremely disagreeable. That early Indiana legislation, for example, had been directed towards people adjudged insane, idiotic, imbecile, feeble-minded, or addicted to rape or cime. There were also the laws concerning immigration into America. These took a decade or two to bring into effect, after modern genetics had been initiated at the start of the century, but they transformed the great open door facing Europe into a series of little doors based largely on nationality. The change was in 1924 but first, and leaning heavily and with admiration upon Ludmerer, it is necessary to see where genetics went wrong and how it became involved with eugenics, another perfectly sound idea that is today tainted in the minds of many almost beyond repair. Britain and the United States behaved very differently over this involvement, and events in both countries must therefore be examined separately.

Highly pertinent to this difference has been the recent name-changing controversy. The American Eugenics Society voted in October 1972 by 108 to 26 that the middle and offending word should be dropped from this name. One month later, and by a similar proportion of 4 to 1, the same organization re-emerged as the Society for the Study of Social Biology. Oddly, and without knowing of the impending American debate, let alone its outcome, the council of Britain's Eugenics Society discussed the same issue. Other names were proposed, such as the Society for the Study of Human Genetics and its Social Consequences, but no new title was welcomed. A substantial majority of the council objected to any change. Therefore the Eugenics Society, the first of its kind in the world, is still the Eugenics Society, despite that word, despite the American change of thought. The British felt that, even though the name is 'not ideal', it 'does not substantially hamper the useful pursuit of the Society's activities'. The Americans, just an ocean away, believed the old name not only inhibited growth in membership but actually hindered the purpose of the Society. It is the history of the eugenics movement in the two countries that has been responsible for the current disparity, and for that singular change in name.

At first it seems reasonable that geneticists everywhere should lend their name and support to any eugenics movement. Genetics is the study of heredity, of the creation of the next generation. Eugenics is, in essence, just a part of genetics, that part devoted to the study and development of improved heredity, of the creation of a better generation. Francis Galton defined it (in 1904) after coining the term as 'the science which deals with all influences that improve the inborn qualities of a race; also with those influences that develop them to the utmost advantage'. Genetics therefore cannot be counter-productive to the requirements of eugenics and about half of the geneticists in the United States became active in the American eugenics movement. What, one might ask, is the harm in that?

To find an answer it is necessary to retreat still further into the past. Throughout the nineteenth century, and in America as much as elsewhere, momentum had been gaining concerning the possible betterment of the human species. For example there had been John Humphrey Noyes. Advocating that the gospel, if accepted, secured freedom from sin, he founded the Perfectionist sect, one group of the so-called Bible Communists or Free-lovers. He further advocated (in 1848) that humanity could be improved by judicious breeding. Charles Darwin's views on evolution were then to shake both sides of the Atlantic, inducing more thought about survival, fitness, and change for the better. Soon came Galton, polymath and Darwin's cousin. Within a few years of *The Origin of Species*, and into the confusion it had caused, he was asserting that 'if a twentieth part of the cost and pains were spent in measures for the improvement of the human race that is spent on the improvement of the breed of horses and cattle, what a galaxy of genius might we not create'.

Subsequently came 'Social Darwinism', the inevitable link between mankind's society and that of every other living organism. It was the direct application of evolutionary ideas expressed in the *Origin* to the human situation, but this meant equating 'the unfit' of the animal kingdom with 'the poor' of mankind. Herbert Spencer, top exponent of this form of naturalism and the 'most immeasurable ass in Christendom' (said Thomas Carlyle), felt the poor should be left to die out. He sold hundreds of thousands of books based upon this and other philosophies, and he also coined 'Survival of the Fitter' (not Fittest). The fact that Galton, Darwin and Spencer were all Englishmen did not detract from their influence in the United States. On the contrary, in a burgeoning country, where success was always crucial and the underdog some form of lesser creature, any talk of fitness and survival was totally pertinent to everyday life.

Then came Mendel, or rather he came again via August Weismann and others. The rediscovery of Mendel's 'particulate inheritance', with this or that gene inexorably dictating this or that character in the next generation, came at a critical time. From sweet-pea colour to blue eyes versus brown eyes and then to every inherited aspect of the human form – it was all heady stuff, this genetics. It was simple, at first. It was comprehensible at first, and if no one delved too deep. It was also all-embracing; each manifestation of human society, from the prevalence of drunkards to every other fault and virtue, came under its aegis, its precise laws of inheritance, its ratios of 1:1, or 3:1, or 9:3:3:1, and on. It fitted the bill for both fact and prejudice. It was convenient for the interwoven web of opinion, belief and experience because it seemed to support all three: with like producing like, breeding always triumphant, races either inferior or superior, and good seed manifesting itself sooner or later with Oliver Twist rising above the lowly environment to which he, and his genes, had been temporarily subjected.

With part prophecy and part hindsight the geneticist William Bateson wrote (in 1908) that:

'So soon as it becomes common knowledge ... that liability to a disease, or the power of resisting its attack, addiction to a particular vice, or to superstition, is due to the presence or absence of a specific ingredient, and finally that these characteristics are transmitted to the offspring according to definite, predictable rules, then man's view of his own nature, his conception of justice, in short his whole outlook of the world, must be profoundly changed.'

In the early years of this century it did become common knowledge. The eugenics movement, instead of being liberally sprinkled with working geneticists, became increasingly and disturbingly the property of other kinds of people. There was David Starr Jordan: 'The destruction of the strong means the perpetuation of the weak. The loss of the bold, dashing and courageous means the rule of the cautious, the timid, the time-serving.' He felt war to be dysgenic – harmful to the race. Others saw it as a gladiatorial contest, and therefore eugenic or beneficial to the race. There was W. E. D. Stokes: 'It is no trouble to breed any kind of man you like, 4-ft men or 7-ft men – or, for instance, all to weigh 60 or 400 lb., just as we breed horses.' Being a horse-breeder of distinction himself he found no difficulty in becoming a eugenicist.

And then there was Madison Grant: on Jews 'their dwarf stature, peculiar mentality, and ruthless concentration on self-interests are being engrafted upon the stock of the nation'; on Indians 'they can scarcely be regarded as human beings' because of their cruelty; and on Negroes 'they are willing followers who ask only to obey and to further the ideals and wishes of the master race'. He was for absolute suspension of all immigration to the US, the deportation of aliens, whoever they were, and the promotion of a eugenic breeding programme. (Presumably it is pointless to reflect upon any Amer-Indian definition of that word alien.)

The geneticists, now learning rather more about their craft, and about the hideous intricacies involved in the creation of one generation by another, did not embrace the eugenics movement so warmly as before, but neither did most of them reject it out of hand. For example there was Charles B. Davenport. 'Proper matings,' he wrote, 'are the greatest means of permanently improving the human race, of saving it from imbecility, poverty, disease and immorality.' Admittedly that was written in 1911, but he continued to promote eugenic ideals, despite his disturbing bedfellows in the movement. So did Edwin G. Conklin, another geneticist, who wrote (in 1928) that 'although our human stock includes some of the most intellectual, moral and progressive people in the world, it includes a disproportionately large number of the worst human types'. Earlier he had written of his belief that 'our civilization, like other civilizations of the past, is showing signs of degeneration and decay, that throughout the world the less intelligent and more selfish elements of society are coming to control government, industry and education, while the best elements are dying out or are losing control'.

Most geneticists were not so much voting against the eugenics movement, as might have been expected, bearing in mind the pronouncements of Jordan, Stokes, Grant and Co., but were increasingly abstaining from it. The eugenic ideal, commendable in that it inspires to improve the human race, was sufficient to cause some geneticists, such as Davenport and Conklin, to continue promoting that ideal, however much they may also and in passing have promoted the gross distortions steadily becoming uppermost in the movement. Not only were the sterilization laws gaining ground, but the nation of immigrants was, for the first time in its history, about to examine its European influx and legislate accordingly. (Immigration from Asia had not been free for a long time. Towards the end of the nineteenth century, and according to official testimony in Washington hearings, it has been stated that: 'The Chinese are inferior to any race God ever made ... I believe the Chinese have no souls, and if they have, they are not worth saving.' The Asian door had never been so wide open as the European door and therefore the new laws concerning Europe were a greater volte-face for a country largely peopled with Europeans.)

Before investigating that particular product of eugenicist vigour, it is timely to wonder how the medical profession was reacting to the onset of genetics and to all the talk about the betterment of the human frame. The short answer is that it was not reacting. Certain doctors had done some of the pioneering work associated with human genetics, such as Karl Land-steiner with blood groups, but the profession itself seemed disinterested. The first required course in genetics for American medical students was instituted at Ohio State University in 1933, or sixty-eight years after Mendel had published his findings and more than thirty years after his experiments had been so triumphantly rediscovered. The major book of those times dealing with human genetics was *Heredity in Man*, written in 1930 by Reginald Ruggles Gates. He was not a medical man himself, but nine chapters of his book were devoted to inherited disorders. Only three years earlier the American doctor most knowledgeable in this field had said: 'To biologists and geneticists it must be little short of amazing that American clinicians have been so little influenced either in work in or thought by the stupendous advances that have been made in our know-ledge of heredity since the turn of the century.' With so much prevalent disease at that time attributable to infection and to the imperfections of society it was presumably simpler to disregard the errors of inheritance, particularly as no remedy was in sight for them.

The situation was therefore ripe for political connivance; there were leaders and there was not much opposition. Genetic thinking had sparked off ideas, well-conceived or otherwise, in the minds of many citizens, and most of the original leading eugenicists, active in the first decade of the century, were still both active and strongly in charge of the movement twenty years later. The initial leavening of geneticists, of men prepared to change their views as fresh facts were unearthed, had withered when their gradual disenchantment caused them to withdraw to their laboratories. At the same time young people felt no inducement to join the movement, and the medical profession, by staying so curiously aloof, did nothing to counterbalance the growing power of the American Eugenics Society with its veteran campaigners. There was a little opposition, but the subsequent events are best explained by a relative lack of it.

The Immigration Restriction Act of 1924 seems to arise as a sudden and dramatic shift in New World policy. So it did, but there had been rumblings beforehand. Grant, yes him again, had been calling immigrants 'the great swamp of human misery and degradation'. President Theodore Roosevelt, that eponymous cause of the teddy bear, was less cosy on racism, referring to the 'yellow peril' and 'race suicide' of his fellow citizens unless they bred more 'native Americans'. Leon F. Whitney, of the Eugenics Society, said the new arrivals were 'tares in the wheat' and 'genuine human weeds'. (One wonders here about false human weeds, and should again be indebted to Ludmerer for having unearthed these and other testaments to human prejudice.)

The official investigation set up (in 1907) to examine the inflow problem was the Dillingham Commission. Immigrants from the Mediterranean regions, it subsequently asserted, were 'biologically inferior' to those from other areas. This official conclusion enabled the racists to make fulsome use of it. There was, for instance, Albert E. Wiggam: 'Investigation proves that an enormous proportion of undesirable citizens are descended from undesirable blood overseas. America's immigration problem is mainly a problem of blood.'

To be fair it was also a problem of mere numbers. In 1912 over 800,000 entered the United States, in 1913 over 1,100,000, in 1914 over 1,200,000. Lack of shipping during the war caused a decrease – only a million arrived throughout the four years of 1915–18 – but the onset of peace meant a resurgence of the old flow and also a revitalization of the former resentment. The time for racial legislation was therefore due. Who knew what manner of Bolsheviks, Catholics, Semitics, anarchists and radicals would be off-loaded from Europe, quite apart from those others who had formerly lived and energetically bred around the Mediterranean?

Suddenly the whole of America was ready for massive group prevention as against mere personal restriction. Laws were already in existence forbidding entry to certain kinds of individual, such as criminals, polygamists and the mentally unsound. There was also that legislation concerning oriental immigration, but the flow across the Pacific had never matched, with or without laws, the tidal wave across the Atlantic – think of over 3,000 arriving every single day throughout 1914.

The sudden national desire was not only to restrict this wave but to choose. Even the Dillingham Commission had only advocated a reading and writing test for all newcomers as a new means for selection, but the eugenicists were to trump this relatively gentle idea, however sweeping its effects for impoverished communities, by increasing talk of 'disharmonious crossings'. Despite proven information about hybrid vigour from the breeding of animals, this devastating theory of disharmony stated that children of a mixed marriage would surely be inferior to both parents. It has often been argued since that the energy of America is largely attributable to its hybridization but, at the time of World War I and immediately thereafter, the new and popular eugenic theory held that American blood would suffer if mixed with foreign blood. The two strains were purported to be incompatible, and the effect on the offspring would therefore be detrimental. Hence, in a rush, the 1921 Immigration Act and then, with more thought but equal determination, the Immigration Restriction Act of 1924.

The plan was to reduce the amount of mixing. It was therefore decided to admit immigrants in proportion to those of their kind already living in the United States. This was racist thinking but was hung, for convenience, on to nationality. The huge twentieth-century influx had been largely from non-Nordic nations, but the legislators decided to disregard that confusion of recent arrivals and to base their laws upon the year 1890. The kinds of people then living in the US were felt to represent the kind of American it was advisable to encourage. Similarly, the type of immigrant who had been arriving since then was, in the main, the kind of person the Act was attempting to discourage.

In short, it was stipulated that annual immigration from each European country should not be greater than 2% of the number of US residents who, in the 1890 census, had given that nation as their place of birth. All subsequent immigration was to be in direct proportion to the flow that had occurred in the decades leading up to 1890. Those years, it need hardly be said, had seen the arrival of many from the Nordic nations, from Scandinavia, Germany, Poland and much of eastern Europe. (The word Nordic is doubly confusing when it refers to, say, a Hungarian, but in the context of 1924 it was somewhat synonymous to non-Mediterranean, and of the two terms one is almost forced to agree that Nordic is preferable.)

The people in charge of the eugenics movement had therefore seen to it that all newcomers would be largely of the same background as, somewhat expectedly, the people in charge of the eugenics movement. Great Britain, for example, had an annual quota of 65,000. It did not make full use of its allowance, partly because there were so many other countries to which Britons were particularly welcome, such as the dominions and the rest of the empire. Italy, with similar associations existing in only a few portions of Africa, and with a booming population, had been making use of America as its major safety valve in the early years of the twentieth century. Suddenly, in 1924, it was restricted to 5,666 migrants a year. Greece, with no imperial connections anywhere and with great unemployment, was allowed – as from that year of curtailment – to send only 308 to the United States annually, or less than one a day. The eugenicists had indeed pulled off their major coup, and its effects were to cause greatest hardships in those very countries with most need of a new world.

For four decades the injustice continued, but the laws were eventually repealed with the Celler Act in 1965. From then on 20,000 was to be the maximum from any one country and, with that proviso, immigrants were to be admitted according to the order in which they applied. The old 'national origins' principle was therefore abandoned and 1890 quietly disappeared, as was its due, into the past. Both the compulsory sterilization and the immigration restriction laws had originally been carried through under the banner of 'eugenics'. Hence the fact that the American Eugenics Society decided, although not until 1972, to drop the offending word and change its title.

Not so in Britain, and there were reasons for the Atlantic difference. The British eugenics movement had certainly acquired some frightening adherents, but many of these had been dominated both by thoughts of class distinction and by the feudal notion that the upper classes did actually inherit something more than just wealth, land, large homes and a feeling of superiority. Such aristocratic ideas were prone to have smaller followings than the large groups with ethnic or racial allegiances in the United States. Secondly, the British were able to achieve a greater degree of good work on human genetics as a science without it becoming confused with eugenics as an ideology, erratic or otherwise.

J. B. S. Haldane, removing the bushel far from any light, said that there were only half a dozen people in the world, including himself, who knew about human genetics. Moreover, they were all British save for the Swedish Gunnar Dahlberg. One of the six men was Lancelot Hogben and his time in South Africa (Professor of Zoology, Capetown, 1926–30) had ended with his personal disgust at its racial policies. He was therefore more sensitive than many to Germany's similar pronouncements and warned that human genetics must be developed as a science without being used for political manœuvre. Consequently, with snobbery as a poor political motivator, and with geneticists taking an active interest in its human angle, the eugenics movement in Britain did not achieve legislative power.

Nevertheless it might have done, and it might have created laws even more punitive than those of Germany. 'Anyone who is hereditarily ailing may be sterilized,' stated the basic German ruling, 'if in the experience of medical science it is with great probability to be expected that his progeny will suffer from severe bodily or mental hereditary disorders.' In other words, the person to be sterilized had himself to be abnormal to some degree. In the 1930s there was a Bill before the British Parliament which, if passed into law, would have been even more savage than the German ruling. It would have permitted the sterilization of 'persons who are deemed likely to transmit a mental defect or a grave physical disability to subsequent generations' whether they had or did not have this grave defect themselves. The Bill was never passed and Britain remained un-sterilized. (As a speculative digression one can wonder if the royal families of Europe, plagued with an undoubted physical disability – haemophilia – could ever have been subject to a sterilization of their blood royal.)

There might have been a more positive eugenics policy in Britain had there been a large Negro population or had there been a similarly pro-Nazi faction as in the United States. The Galton Lectures, delivered annually in London under the Society's auspices, show the range of interest, somewhat disturbingly. 'What Nations and Classes will Prevail?' asked Dean Inge (in 1919). F. C. S. Schiller spoke (in 1925) on 'The Ruin of Rome and its Lessons for Us'; the Bishop of Birmingham (in 1926) on 'Some Reflections on Eugenics and Religion'; C. J. Bond (in 1928) on 'Causes of Racial Decay'; and E. J. Lidbetter (in 1932) on 'The Social Problem Group as illustrated by a Series of East London Pedigrees'. The movement amused itself, and studied this and that, but never became a threat. Finally, as Ludmerer puts it, 'the more astute English geneticists saw a bit earlier than the Americans that the fate of human genetics did not have to be wrapped up with that of eugenics'.

The word, in short, never became so tarnished within the United Kingdom. Compulsory sterilization was never implemented. The selec-tive immigration laws, imposed in the 1960s after Britain had for the first time become a net-importing country of people, following a major inflow from the West Indies, certainly had racial tendencies but were not enacted under the name of eugenics. Nevertheless the activities in America, Germany and various other nations have left marks, some totally valid, some merely tedious. 'I am writing a book on human breeding,' I said to a friend. 'Fascist,' he replied.

'Well, we all know what the Germans did,' we all say. In books, given half a chance, there is brief and uninformative reference to the famous German attempts at 'breeding supermen', to 'vile Nazi practices'. In articles there are similar swingeing statements, rich in chastisement, short on information. Consequently, or so I suspect, we have ended up knowing that something went on but knowing little more than that. How many of us, for example, have even heard of Lebensborn?

It is necessary to go back to *Mein Kampf*, Adolf Hitler's work that ought to have been required reading outside Germany just as it was forcibly decreed within the Third Reich. An Englishman wrote in 1936, after Hitler's NSDAP had been governing for three years, that 'our generation has become familiar with politicians who frame radical programmes in youth and who repudiate them as soon as they are overtaken by the responsibilities of power: not so with Hitler'. The writer was right. Within those three years, and during the subsequent nine of Nazi history, the original doctrines vehemently expressed in *Mein Kampf* were put into savage effect. The racial and eugenic statements in that book formed no exception to this rule, however exceptional Hitler was as a politician in actually carrying out his printed promises. Here are some quotations.

'Breeding between two creatures not quite equally high yields a product which is a medium between the levels of the two parents. This means that the offspring will indeed be higher than the racially lower half of the parental pair, but not so high as the higher. Consequently, in the struggle against this higher species, it will later on succumb.'

'The prerequisite for improvement of the species lies not in the union of the superior and the inferior, but in the complete victory of the former. The stronger must dominate and not mix with the weaker, and thereby sacrifice its own greatness. Only the born weakling can feel this to be cruel.'

'The mixing of higher and lower races is clearly against the intent of Nature and involves the extinction of the higher Aryan race . . . Wherever Aryan blood has mixed with that of lower peoples the result has been the end of the bearers of culture.'

'Human culture and civilization on this earth are inseparably bound up with the existence of the Aryan. By this extinction or decline the dark veils of an uncultured age will descend once more.'

'Everything we admire today on this earth – sciences and art, technique and inventions – is the creative product of but a few peoples and perhaps originally of *one* race.'

Nowhere does Hitler define Aryan; nowhere does he define race. Nevertheless he could never be accused of being circumspect about his genetic beliefs. Elsewhere in the book he is less philosophic, more direct about his aims:

'The racial state (Völkische Staat) must make race the centre of the common life. It has to preserve the purity of the race. The child must be declared the most precious good of the nation. The state must see to it that only those who are healthy produce children.'

'Never yet was a state founded by peaceful economic activity, but always alone by the instinct of the self-preservation of the species whether this be of the heroic-virtue type or of subtle cunning. The outcome of the first type are Aryan states of work and culture; of the second Jewish parasitic colonies.'

'A racial state will have to raise marriage from the level of a permanent racial shame and to consecrate it as an institution, which is established to bring forth images of the Lord instead of abortions between man and the ape.'

'There is a price to be paid. The Germanization of the future must admit new soil alone. To bastardize is to sin against the Racial Holy Ghost.'

The warning was clear. The Jewish people were in extreme jeopardy, and so were all others on those new lands who might appear insufficiently Aryan. In a sense nothing that Hitler wrote was new. The undefined Aryan had been given birth, or rebirth, in the writing and thoughts of Nietzsche, Treitschke, Houston Stewart Chamberlain, Friedrich von Bernhardi, Wilhelm Marr and many others, but Hitler gave additional substance to their joint offspring, to their combined concept of an Aryan. Actually Nietzsche's 'blond beast' was founded more upon the Russian peasant than his German counterpart, but this fact became submerged beneath the enthusiasm to equate Aryan with German. Nevertheless, and for the record, he did say: 'A thinker who has at heart the future of Europe will in all his perspectives concerning the future calculate upon the Jews and the Russians as above all the surest and likeliest factors in the great play and battle of forces.'

What Hitler subsequently did to prevent reproduction of the Jewish people will be known for all time; so too his extermination of yet greater numbers formerly living on the new and conquered lands. What he also did to prevent progeny descending from the tens of thousands of Germans considered unfit for reproduction has already been described, but his actions actually attempting to promote Aryan stock are the concern of the next few pages.

It was not so much Adolf Hitler as Heinrich Himmler who was the arbiter of this particular variety of positive eugenics. Negative eugenics is the prevention of breeding of certain individuals or types, whether by death or by law. The positive side is the encouragement, whether ill-advised or not, of certain individuals or types to procreate more than might otherwise be likely. On this subject Himmler had been much influenced in the 1920s by Richard Walther Darré, an Argentine German, educated at King's College School, Wimbledon. Darré, says Heinz Höhne (in *The Order of the Death's Head*), preached that the agricultural problem was not primarily economic but one of blood, of good or bad peasant blood. Later, when Himmler had formed his ss, he placed one of the five departments under Darré's charge. This was the notorious Race and Re-settlement Office (Rasse und Siedlungshauptamt – RUSHA) and he was to combine this job with that of Minister of Agriculture, a curious combination it would seem until one remembers the 'blood and soil' mythology that had brought Darré into favour in the first place.

RUSHA was given the task of 'purifying and standardizing the ss'. It recruited Bruno K. Schultz to draw up a scale of racial characteristics and he responded promptly with five groups: Pure Nordic; Predominantly Nordic; Harmonious bastard with slight Alpine, Dinaric or Mediterranean traits; Bastards of predominantly eastern Baltic or Alpine origin; Bastards of extra-European origin.

That Nordic ideal was no vague entity but most precisely delineated. Here, for example, giving to this alleged grouping of mankind both credit and debit where they are felt to be due, is a somewhat lengthy but exceptionally firm definition, written by E. Fischer in 1923:

'The mental endowments of the Nordic race are great energy and industry, vigorous imagination and high intelligence. Conjoint with these are foresight, organizing ability and artistic capacity (this being least marked in respect of music); and also the unfavourable qualities of strong individualism, a lack of community sense and of unwillingness to obey orders, a certain one-sidedness and an undue inclination towards imagination and flights of fancy, a dislike for steady and quiet work; while as additional qualities may be mentioned a considerable expansive force, a power of devotion to a plan or idea, an adequate capacity for instilling an idea into others and a small inclination for adopting the idea of others – in a word significant powers of suggestion and comparatively little suggestibility. It is obvious that when circumstances are favourable persons richly endowed with these gifts are likely to become leaders, inventors, artists, judges and organizers.'

Only the first three of Bruno Schultz's racial classifications, ranging from the Nordic to the harmonious bastards, could be chosen to join the black ranks of the ss. Other physical traits were also examined because Hitler, in particular, had an aversion to those 'who might be tall but were in some way disproportioned'. Schultz again felt equal to the task and listed nine kinds of physique; only the first four passed his test. He was never asked to draw up any classification concerning intelligence. The right shape, size and appearance plus an ability not to behave like an underling corresponded, as Himmler put it, to 'the ideal which we set ourselves'.

All of this was happening long before Hitler's party was voted into power. For example, on 31 December 1931 the Reichsführer ss ordered that members of his organization should be retained 'only if the necessary conditions of race and stock were fulfilled, and after approval by RUSHA and himself'. The ss were therefore being groomed not so much as an élite Nazi corps but as progenitors of the future. Strictures were further strengthened in 1936 when Himmler decreed that his men should marry when aged between twenty-five and thirty, that they and their mates should undergo further tests of physique and origin and RUSHA would then decide if they could be entered in the Clan Book, a record of all those who fell within the compass of Himmler's ideal.

At the same time the German people as a whole were being encouraged to breed. Even the word family, the 'germ cell of our nation', was reserved for parents with four or more offspring. On 12 August, Hitler's mother's birthday, bronze, silver and gold 'honour' crosses were awarded to mothers of 4, 6 and 8 children respectively. Youth organizations were instructed to salute these bemedalled mothers, and all men had to give up their seats to any woman with child, both before and after it had been born. The status of the mother was said to be 'equal to that of soldiers in the thunder of battle' and mothercult acquired its slogan: 'I have donated a child to the Führer.'

The time was also considered ripe for the creation of Lebensborn, the Spring of Life. This organization, founded on 12 December 1935 by the Reichsführer ss, was incorporated within the genealogical office of RUSHA. In many ways its various aims, save for the qualifications, were of some beneficent welfare organization: supporting large families – of good racial stock; taking care of expectant mothers – of good stock provided the father is also sound; and of helping the children. In a report submitted to the Nuremberg trials (of 1946) it was stated that the organization had dedicated itself primarily to the fight against abortion, extremely high in pre-war Germany, and had raised the unmarried mother to a place due to her, for 'in the eyes of the racial community' she is a mother no less so than the married kind. However, Lebensborn did only 'embrace' children who were acceptable from 'a racial and biological point of view'. The rest were handed over to the NSV, the state welfare organization.

It was because of its restrictive characteristics that Lebensborn e.V., an incorporate society, became wedded to the ss. After all, each member of the Schutz-Staffel had been selected for his physical merits, and would therefore serve well as a reproducer of those merits. More German children of all kinds were needed, particularly when it was realized that all those Slavs in the east were breeding faster still, but children of good stock were needed most of all. Besides, a war was approaching that would not only kill many young men but also prevent a similar number of young women from becoming mothers.

Himmler, fearing for the loss of his chosen race, said that no one within the select organization should be so selfish as to lay down his life for the Fatherland without first creating a child as his replacement. These progeny should be of the same high quality – Edelprodukte – as their fathers. Therefore the mothers must be similar to the fathers, insofar as this was possible, in order to ensure genetically true offspring.

The simplicity of such reasoning will offend geneticists, but again and again in this book it should be remembered that Mendel's work, encountered with such enthusiasm in 1900, described only the simplest forms of inheritance and was consistently misunderstood. His experiments showed the effects of only one or two pairs of genes whereas most of inheritance is a plethora of genetic effects, with great numbers of genes each exerting their own variety of influences. To conduct trials with flowers in a Brno garden is one thing; to expect similarly positive answers from the breeding of humans, whether for ss characteristics or not, is quite another. The same rules cannot apply because the same degree of knowledge does not apply. It was genuinely thought, and not just by Hitler and Himmler, that Mendelian simplicity could be immediately applied to the human system. Hence the belief, in fact the certainty, that an ss man coupled with a suitably chosen mate would result in future raw material of precisely the same calibre.

Be all that as it may, the Lebensborn organization, laudable in its welfare attributes, became an additional means whereby ss men could leave a child as their replacement. It provided a meeting place. It provided a mate. And it guaranteed to provide for all the needs of any subsequent offspring. At a stroke, and from the ss point of view, it seemed to satisfy every requirement.

In Louis Hagen's *Follow My Leader* he tells the story of Hildegard Koch, daughter of a baker, golden-haired and well made, who became involved with Lebensborn. Her tale is a good example of the system at work. Born in 1918 she lived in Berlin. Her father was in the sa, the brown Guard, and she too became an enthusiast for Hitler, joining the Bund Deutscher Mädchen. Her BDM leader said the chief aim of the German woman should be to produce healthy stock, and Hildegard was a perfect example of the Nordic type with 'long legs, a long trunk, and broad hips built for child-bearing'. She listened but went to work on a farm, and then fell in love with a local aristocrat who happened to be anti-Nazi. Following the inevitable break-up of this relationship, and during her subsequent disarray, she met again her old Gau leader who advised a Lebensborn home as the cure for her ills. Why not give the Führer a child?

The home was in an old castle, near Tegernsee in Bavaria. There were forty girls, all young, and none knew the names or backgrounds of the others, but each had a certificate of Aryan ancestry as far back as their great-grandparents. 'There had never,' reported Hildegard, 'been the smell of a Jew in her family.' Each girl had a private room, and there was also a library, a music room, a small cinema, good food to eat, a large lake near by and horses to ride. Everything was in the charge of an ss doctor but there was also an ss woman who instructed the girls in their duty. To help in the Nordification of the nation, she said, was an honourable undertaking and each person should be proud of it. Then, following further medical examination, the girls had to renounce all children they would have in that place, for these offspring would be needed by the state and would be taken to special settlements for eventual intermarriage.

The ss men also staying at the home were then introduced to them. They were all tall and strong – the girls each had to be over 5 ft – with blue eyes and blond hair. After one week the girls were asked to choose their partners, but then had to wait until the tenth day after the start of their menstruation. Following one more medical examination each girl was permitted to receive the chosen consort in her own room. 'As both the father of my child and I believed completely in the importance of what we were doing, we had no shame or inhibitions of any kind,' said Fräulein Koch. For three alternate evenings in that week she slept with him, and for the other three her man was away feeling no shame or inhibition with another girl. Neither girl ever learned the name of this temporary partner in their lives.

When pregnancy was proven Hildegard was permitted to go straight to a maternity home. It was 'nice but dull'. The burgeoning inmates were all called 'Frau' because the descendants they were giving to the nation 'were considered more important than the accident of marriage'. Hildegard's baby arrived at term. She breast-fed him for two weeks and then handed the boy over to the state. It had not been an easy birth, 'for no good German would think of artificial aids such as the pain-killing injections they have in degenerate western democracies'.

It is not stated whether it was an easy parting between mother and child. However, she was thanked for her co-operation and invited back to Lebensborn the following year – 'the more children she had the better'. In fact, she became Frau Trutz instead, having found a man to marry, a Gestapo NCO. They both submitted to the RUSHA investigation as to their suitability for marriage. Eventually they were awarded their Ahnenpass – a certificate of pure breeding, or Ancestors' Pass – and were then told to get on with producing the ideal of five children before Hildegard was thirty-five. However, when Ernst discovered that his wife had had a child at Lebensborn, he pronounced extreme and reactionary displeasure at his wife's former behaviour, but he remained loyal to the cause – and to Hildegard.

Whatever the intentions of Himmler and the other instigators of Lebensborn, the society was both slow to start and then overtaken by events. Like Trutz, other members of the ss had fixed ideas about marriage and the role of women in their society. Many of them objected to Himmler's marriage rules and preferred the traditional ceremonies. If discovered the offenders were expelled from the ss but in 1940, when needs had changed, they were welcomed back provided – as was the constant refrain – they were still biologically and genealogically acceptable with that minimum requirement of being a harmonious bastard. Not every ss man was as obliging as Hildegard's alternating lover, and not every German girl was so selfless as Hildegard, but by 1939 Lebensborn was able to state that '832 valuable unmarried German women had decided, in spite of the greatest hardship, to donate a child to the nation'.

The organization was further confused when it was realized, after some of Hitler's new lands had been added to the Reich, that many of their peoples were just as Aryan as, if not more so than, the Germans. Poland was full of blue-eyed, fair-haired youngsters. So, even more, was Norway. And so was Russia. Many children were therefore exported to Germany, referred to as orphans whatever their actual status, examined by RUSHA, and then transferred, if suitable, to Lebensborn homes. The German children, who had been donated with such selflessness, were soon swamped by greater numbers of other children forcibly imported with infinitely greater hardship. Some were not even imported and Lebensborn homes were initiated in Norway, that most coveted breeding ground of the Nordic race.

By the war's end no one really knew, or knows yet, to what extent Himmler's racial brainchild had grown. Heinz Höhne reports the existence of eight maternal and three infant homes in Germany itself, but adds that 11,000 offspring were under Lebensborn's management. (As each birth, assessed in terms of its subsequent labour value, was reckoned to be worth 100,000 marks, the founders of the organization had exulted at their creation of German national assets worth many millions.) The society's actual membership – those people, largely ss, who had supported it financially – only expanded modestly during the war, from 15,777 in 1939 to 17,000 by 1945. One also presumes that the actual breeding department of Lebensborn, with soldiers living in lakeside homes entertaining nubile Nazis for the long-term benefits resulting from their joint labour, had to be shelved while the short-term needs of total war were first satisfied.

At all events the Spring of Life existed for $8\frac{1}{2}$ years, for roughly one-third of one human generation. As a genetic experiment it has nothing or little to teach us. As an ideology it was unpleasant, but the mere coercion of certain men and women to try to reproduce their kind was no crime. The removal of other Aryan children from conquered homes and families quite eclipses any dislike we might feel for the selection and governmental approval of certain matings or for the compulsory orphanhood that it involved. The attempt to breed from a chosen few was without consequence and the few products have now gone back into the melting pot — along with the rest of us.

CHAPTER 15

Eugenics present

Human improvement – Eugenic methods – Unnatural selection
– The Edinburgh Charter of Genetic Rights – Dysgenic
inertia – Positive or negative – Current thinking

*'It is clear that the general quality of the world's population is not
very high, is beginning to deteriorate, and should and could be
improved'*

JULIAN HUXLEY

*'How many women would be eager and proud to bear and rear a
child of Lenin or Darwin! Is it not obvious that restraint, rather
than compulsion, would be called for?'*

HERMANN MULLER (in 1935)

I like the story of the doctor who, confronted by the need for fresh
spermatozoa to effect artificial insemination, and knowing that quality
sperm devoid of blatant genetic defects would be most suitable, suddenly
realized how the problem could best be solved. The answer was, so to
speak, readily to hand. What finer and more convenient sperm than his
own? The lady was inseminated and a child, somewhat in his own image,
was subsequently born. Presumably, as his procedure possessed obvious
advantages of cheapness, availability and self-induced flattery, it could be
utilized more than once. Furthermore, as insemination is being practised
more and more frequently, such a doctor would have increasing recourse
to his own resources. It might therefore be pleasing to sit in his waiting
room, some years and many conceptions later, and perhaps to observe
the multitudinous fruits of his simple idea. A man so fertile both in
imagination and body could reflect happily, like some Hebrew patriarch,
upon the inspiring numbers of his begotten progeny.

The story also indicates, which is its principal purpose, the kernel of
this chapter. Who will decide whether positive eugenics, the deliberate
encouragement of certain forms of offspring, is to occur? If it does, who
then will donate the sperm? And if a certain kind of sperm is customarily

used, what effect, and how speedily, will this have upon succeeding generations? Does it matter that more harmful genes are being handed down to the next generation than were handed down to us? Does it matter if harmful mutations are less lethal than they used to be and are therefore accumulating in the population?

Muller achieved his initial fame and his Nobel prize by proving that X-rays cause mutations. His reputation on that score helped to provide a wider audience for his eugenist views and he once wrote that, if present trends continue, the whole world will become a hospital with 'even the best of us only being ambulatory patients in it'. Ought steps, therefore, to be taken to reduce the transmission of harmful inheritance? Mankind is undoubtedly furthering the science of genetics, but should not genetics be used more actively to improve mankind? Even if we fail to agree over most definitions of an improved human being, resenting those proposed by Laughlin, Grant, Jordan, Stokes, Himmler, Schultz, and Darré in the previous chapter, we do not disagree that humans could be improved. There is no glorious magic in the wide-ranging bunch we happen to be and which we reproduce by the method that evolution has, most haphazardly, handed down to us. It is good to be variable but there is no good in some of the extremities of our variability. Of what purpose is brainlessness? Of what point is death in infancy? For those who argue that all painful incidents are blessings, most heavily disguised and aiding our immortal souls, there is no parallel argument that today's proportion of malfunctioning bodies is somehow optimal. Are there a satisfactory number of diabetics? Was it better when, before 1922, they died for lack of insulin? Or is it better now that their survival is leading to a greater number of diabetic descendants?

As for intelligence, do we wish for more of it? We can resent the ill-fortune of the grossly stupid but to what extent do we long for more of us at the other end of that scale, for more Einsteins and Newtons? Man has already proved himself sufficiently clever to endanger his own existence, and such geniuses have helped to place us, so to speak, ahead of ourselves. Ought the basic norm of intellectual capacity to be raised so the average man of tomorrow is the clever man of today? Or perhaps intelligence, so overrated in all those years of education, is still overrated by the rest of us. What association has it with happiness, with kindness, with co-operative living, with any other virtue but its own?

There is concern that mankind should tinker with anything so crucial as his own generation. Certainly he has killed, and maimed, and tortured, and he has all but slaughtered the planet; yet he has left his personal pro-creation to the gods and to chance. Will this always be so? Jean Rostand, the French author, has written that 'one is fully conscious of the formid-able difficulties raised by the idea of an evolution controlled and directed by man'. Indeed, but as an item for concern, the German attempt at mating like with like for the betterment of the race was relatively harm-less. If it was a crime it cannot be compared with the excesses perpetuated in virtually every other sphere of Nazi endeavour. Besides, as is known, mankind is already directing its evolution by the inconspicuous means of taxes and child allowances, by religious and political argument, by laws and customs. Should not the direction be more conspicuous, the intent better known, the results more ably appreciated?

Eugenics had such a bad start, as expressed in the last chapter, that even a mention of the subject is often disliked. Some of us are totally unwilling to exchange our prejudices for a realization that noble ideals were also enmeshed within eugenics, and that eugenicists were and are disturbed about the future of our species. 'Look what the Germans did' is, these days, an inadequate answer to every question involving the genetic betterment of mankind. Look what we might do, now that we know rather more and have the mistakes of the past behind us, is surely more valid.

As a final introductory point this current age is not the first to have con-sidered the topic. Plato, in his *Republic*, wrote that the aim of an ideal city is

'... to make marriages as healthy as possible, that is to say as advan-tageous as they can be to the state, and to that end we must learn from the breeders of hunting dogs and birds of prey ... In the same way, if we want to prevent the human race from degenerating, we shall take care to encourage union between the better specimens of both sexes, and to limit that of the worse.'

A century earlier Theognis of Megara had written: 'One would not dream of buying cattle without thoroughly examining them, or a horse without knowing whether he came of good stock; yet we see an excellent citizen being given to wife some wretched woman, daughter of a worthless father.' Campanella, in *The City of the Sun*, echoed precisely the same point in the seventeenth century AD. His narrator says: 'Indeed they laugh at us who exhibit a studious care for our breed of horses and dogs, but neglect the breeding of human beings.' The utopia of his book had a Ministry of Love whose business it was to watch over marriage and procreation. 'It is no more difficult,' wrote Robert le Jeune at the start of the nineteenth century, 'to have intelligent children than to have an Arab horse, a short-legged Basset hound or a pure-bred canary.' There have also been detractors. 'Let us hope,' said Georges Duhamel, 'that the humanity of the future, governed by the new sciences, will engender as many outstanding men as it has produced hitherto at random and in ignorance.'

Now to the eugenic methods that might either be incorporated or more strongly promoted within our society, this current 'humanity of the future'. They can be grouped under five headings.

Firstly, artificial insemination. Whether another man's sperm are used to fertilize a woman, her husband's sperm having proved inadequate, or whether another woman's eggs are used, someone has to choose the donors of these particular germ cells. A form of artificial selection is being used, and will be used increasingly in the future as more means are found for stepping up the fertility of otherwise infertile unions. This subject is so important, both in its implications and in its possibilities, that it forms the basis of the next chapter.

Secondly and thirdly are genetic surgery and cloning. Both are entirely distinct as techniques but they can be lumped together because their importance is more distant. Atomic engineering with the molecules of inheritance may prove to be feasible, and may even be able to improve an otherwise defective gene, but is it likely to become common practice? The world is rich, as it knows only too well, with entirely effective sperm and eggs. The manipulation of a few of those that are at fault may help to solve a few personal tragedies and also lessen the accumulation of harmful genes, but will do little to direct evolution. The possibility of creating beneficial mutations, and of doing this to an appreciable extent, is even more darkly in the future.

As for cloning, what manner of state and what kind of society would either promote or tolerate the massive duplication of human beings? The time, one feels, for that kind of rejection of the individual spirit lies in the past. Rulers are not so absolute today as they used to be; neither are their subjects quite so subjected. The future has to be viewed with extraordinary pessimism to expect that some of the oppression of the past, which cared not a whit for the rights of each or any man, would be multiplied a hundredfold at some date yet to come. Perhaps the crystal is too cloudy for any period beyond our own immediate lifetime but no clone will arise within that span. Or so I believe. And so, one hopes most dreadfully, say all of us.

Fourthly is the category that comes nearest to the German idea. The state, which exerts only modest control at present over who shall mate with whom and how many children they will have, could well become more powerful. It now vetoes particularly inbred partnerships; it could well extend that ban to cover a wider range of near relationships. It now exercises financial influence over every couple's breeding performances; it could do more, blessing with cash those that are favoured by some genetic board of control. (Incidentally, it may be unfair to use the word state, instead of community, or government, or the law, or society, but every word is loaded in some direction and 'the state' has, I feel, a fair and brutal honesty about it.) Each state does both prevent and influence; each could do more. The state also pretends to have no genetic policy; nevertheless any welfare programme, any change to the style of living and every tax rebate does have evolutionary implications. Any government would welcome a reduction in genetic abnormality or a gain in fitness. It would welcome fewer hopeless cases and be grateful for any lessening in the price it has to pay (medical care, extra education, days off work) for every defect that might have been avoided by more judicious reproductive methods. Control by authority is insidious at present; it could become yet more insidious or further into the open. It is up to us to choose.

Fifthly is the eugenist dream, that the best will always reproduce the most. The hypothesis has been around for a long time, but Frederick Osborn in the 1930s had much to do with its current formulation: that a welfare state could be so arranged that it would bring about eugenic selection by the environment. There were two prerequisites to his idea: that the environment of our lives should be improved – extreme poverty, ill health and ignorance must all go – and there should be total freedom of parenthood via contraception. With those two points settled it would be occasion to introduce the third factor, the one which 'will tend to encourage births among parents most responsive to the possibilities of their environment, and to diminish births among those least responsive, thus bringing about a process of eugenic selection through variations in size of family'.

Certainly it appears simple. After all, we each have to live in a particular environment, partly of our own and partly of nature's making, and why promote those of us least suited to it? The scheme does away with countless decisions, about the kind of intelligence that is best, the kind of physique, the sort of personality, the type of individual. Osborn's responsiveness to the environment is, more or less, a euphemism for success.

Much of the world measures success in cash terms, or in power, or by some form of social ladder, but envy or outright disgust can cause many of us to resent any measurements. Nevertheless we cannot deny that we are not all equally suited to the environments that are imposed upon us. Some people, for example, were successful schoolchildren: they flourished in that particular world; they benefited from it and, more like than not, gave to it as well. Some fit so well into the army that they are at a loss when peace breaks out. Some welcome heat or cold while others wilt or freeze, faring less well physically or mentally or both. The misfits are only too aware of their predicament. They observe those who fit into neat niches, but no amount of coercion will make the misfits fit so well. They may have other abilities, more suitable to another century or another planet, but they can be out of tune with this one. A few of us may believe a few of them to be the salt of the earth but nature would think them unsuited, less successful and of lesser value for fathering the next generation.

In a sense the Osborn hypothesis is no more than a return to the law of the wild. Evolution takes little heed of many enterprising sidelines, for it rigorously selects those most suited for the particular place they live in. In the outside world this selection is a natural business, but mankind's domain (contraception, money, stimulants, propaganda) is artificial. Man-made and equally artificial contrivances will therefore have to be employed (contraception, money, stimulants, propaganda) to make certain that the most successful are the most prolific. Something of the sort must have happened naturally in the cave or in the Neolithic dwelling; the man with the ability to produce most food would, probably, keep most children alive. Osborn's philosophy is to return to that state but within the framework of a modern society.

It was shortly after he had clarified the eugenic theory, and shortly before World War II started, that a related and relevant document was drawn up which came to be known as the 'geneticist manifesto'. The original signatories formed an extremely distinguished group, about as far removed from Laughlin and Darré as imaginable in their approach to the subject; namely, F. A. E. Crew, J. B. S. Haldane, S. C. Harland, L. T. Hogben, J. S. Huxley, H. J. Muller and J. Needham. The document, also known as the Edinburgh Charter of the Genetic Rights of Man, was drawn up during the Seventh International Congress of Genetics held in that city and only adjourned the day before Hitler's invasion of Poland.

By that time German perpetrations being performed in the name of genetics were well known, at least to geneticists, and the vanguard of genetic opinion felt it timely and appropriate to draw up a firm statement of its joint belief.

There were seven major declarations which can, to some extent, be summarized:

1. The effective genetic improvement of mankind (as Osborn had stated) is dependent upon major changes in social conditions that would create approximately equal opportunities for all.

2. Changes would also be necessary in the current economical and political conditions which foster antagonism between different peoples, nations and races.

3. The removal of race prejudices and of the unscientific doctrine that good or bad genes are the monopoly of particular groups will not be possible before the conditions which make for war and economic exploitation have been eliminated. This requires some sort of effective federation of the whole world.

4. There has to be legalization, development and dissemination of birth control systems, both negative and positive, that can be put into effect at all stages of the reproductive process. At the same time there has to be development of social consciousness and responsibility in regard to the production of children. The superstitious attitude towards sex and reproduction has to be replaced by a scientific and social attitude. It will then be regarded as 'an honour and a privilege, if not a duty, for a mother, married or unmarried, or for a couple, to have the best children possible, both in respect of their upbringing and of their genetic endowment, even where the latter would mean an artificial – though always voluntary – control over the processes of parentage'.

5. The intrinsic (genetic) characteristics of each generation can only become better as a result of selection. Those persons with a better genetic equipment have to produce more offspring than the rest, either consciously or as an automatic result of their way of life. With modern living an automatic selection is less likely than in the past; therefore some guidance of selection is called for. To make this possible 'the population must first appreciate the force of the above principles and the social value which a wisely guided selection would have'.

6. The most important genetic objectives are the improvement of those characteristics which make (a) for health, (b) for the complex called intelligence and (c) for those temperamental qualities which favour fellow feeling and social behaviour rather than those which make for personal 'success', as success is usually understood at present.

7. Much more than prevention of genetic deterioration is possible. The raising of the average level of the population nearly to that of the highest now existing in isolated individuals could be the aim. Genetically this achievement would be physically possible within a comparatively small number of generations. 'Thus everyone might look upon "genius", combined of course with stability, as his birthright. And, as the course of evolution shows, this would represent no final stage at all, but only an earnest of still further progress in the future.'

The Edinburgh meeting was disbanded, the war began and 50 million were to die in a conflict not only caused by 'antagonism between different peoples' but abetted by the 'doctrine that good or bad genes are the monopoly of particular groups'. Muller wrote, when that war was over and other prospects were looming, that it would not be amiss to recall the 'statement of fundamentals, drawn up and subscribed to at a very solemn time, by some of the leaders of genetic thought'. Apart from those who created the original document there were others who later wished to attach their name to it. They were G. P. Child, P. R. David, G. Dahlberg, Th. Dobzhansky, R. A. Emerson, C. Gordon, John Hammond, C. L. Huskins, W. Landauer, H. H. Plough, B. Price, J. Schultz, A. G. Steinberg and C. H. Waddington. From this list of names and from the original seven it should be manifestly clear that the basic hope of genetics for the betterment of mankind is neither some crazy dictatorial whim nor a malicious endeavour forever to be brushed aside by prejudice arising from the earlier mismanagement of the eugenic idea.

To a greater or lesser degree every one of the signatories continued to promote the notions embedded within that manifesto, and not the least was Julian Huxley. In his *Essays of a Humanist* (published in 1964) he hammers away at the same theme. Of course mankind's organization has been progressively improved, he argues, from a single-celled to a multi-cellular form, from a three-layered type with many organ systems to a chordate with gill-slits, and to a jawless and limbless invertebrate, to a fish, to an amphibian, to a reptile, to a mammalian insectivore, then to a lemuroid, then a monkey, then an ape-like creature 'and finally to a full human organism, big-brained and capable of true speech'.

There is also, he adds, the other side of the coin. Man is a highly imperfect creature. He carries a heavy burden of genetic defects. As a psychosocial organism he has not undergone much improvement. Indeed 'he is still very much an unfinished type, who clearly has actualized only a small fraction of his human potentialities'. Worse than that, he may be slipping backwards, with genetic deterioration being rendered 'probable' by his social set-up and 'definite' by the increased levels of man-inspired radiation. The first task therefore is to stop the harm, to reduce that extra radiation to a minimum, to discourage defective kinds of people from breeding, to lessen human 'over-multiplication in general, and the high differential fertility of various regions, nations and classes in particular'. Only then can man 'proceed to the much more positive task of positive improvement'.

The conclusion, to him, is straightforward: 'It is clear that for any major advance in national and international efficiency we cannot depend on haphazard tinkering with social or political symptoms or *ad hoc* patching up of the world's political machinery, or even on improving general education, but must rely increasingly on raising the genetic level of man's intellectual and practical abilities.' The same inevitable conclusion appears again and again, and with increasing urgency: 'To be effective such non-natural (human) selection must be conscious, purposeful and planned. And since the tempo of cultural evolution is many thousands of times faster than that of biological transformation, it must operate at a far higher speed than natural selection if it is to prevent disaster, let alone produce improvement.'

In short, as Ernst Mayr has written, there are two attitudes we can take with reference to the future. 'We can assume the attitude of the watchers of a Greek tragedy – without raising a finger let the play drift inexorably toward its blood-stained conclusion. Or we can behave like utopians, to a greater or lesser degree, and propose measures that will better the fate of mankind and hopefully better mankind itself.' To quote Muller once again (and he, after all, has been a patriarch of this particular science), the guidance of the heredity of man's own nature 'must for a long time to come be the hoped-for ultimate goal of applied genetics, far overshadowing any of its other possible attainments'.

Despite Muller and all those of the Edinburgh manifesto, despite the firm logic of what they say and even though there is great awareness of the

good that can be done via genetics, not least in the prevention of accumulating harm, the traditional reaction is still to reject the possibilities. To
do nothing is, so it is said, to risk nothing, to court no disaster, to leave
well alone. The guiding spirit of such cautious inactivity must be that
inspector of inventions who turned down every device submitted to him,
arguing that most inventions prove to be valueless and therefore it is
wisest to reject them all. Why not maintain the status quo?

Unfortunately, as we now know only too well, there is no such status
in the genetic composition of populations. There is always change. There
are inevitable mutations, almost always for the bad. There will be more of
them in an increasingly radioactive world. There is continual selection.
One man and one woman have ten children; another man and another
woman have one. At the individual level their contributions to the
gene pool of the next generation are in that ratio. At the population level
certain groups, or nations, or forms of society are contributing more than
others. There is no status quo in this living world; instead there is perpetual change from each generation to the next. The only real question is to
what extent man will impose his positive intent as opposed to his random
behaviour upon that eternal alteration.

Intelligence is a fair example of both potentialities. It is frequently the
worst example to choose because the merest mention of the word can lead
to greater stupidities being proclaimed than with any other aspect of
inheritance, but there is considerable evidence that the most intelligent
among us are not producing the most children, their fair share of the next
generation. There is also evidence that the unintelligent, as opposed, one
hastens to add, to the poor, are creating more than their share. Therefore, if these affirmations are correct, a lower intellectual level is being
selected for at the expense of a higher. If unchecked, and if true, our most
superior ability over the animals will indeed diminish. As Haldane wrote
(for Britain): 'If the existing differences in fertility of the social classes
continue, we may expect a slow decline perhaps of 1 or 2% per generation
in the mean IQ of the country. That is, on the whole, deplorable.'

There could also be a positive side. The average IQ is, by definition,
100. If every child of the next generation were born to average females
but with fathers of IQ 140 the result would be an average much higher by
today's standards. One more such generation and the average would be
higher still. By then fathers of IQ 140 would be relatively commonplace
and it would be necessary to provide fathers of 160 or so to maintain the

upward rise in intellectual ability. Such a wholehearted attack on the problem is unlikely in our immediate society but it indicates the possibilities. As the British geneticist C. O. Carter has written: 'The social consequences of a change of average intelligence of this size would be enormous.' They would also be enormous if the change were to go the other way.

However, even more modest assaults have intriguing implications. It has been calculated that if the average genetic IQ of a population is raised by $1\frac{1}{2}\%$ there will be a 50% increase in the number of individuals with an IQ of 160 or more. Intelligence is not the only ability of man's brain – think of imagination, or inspiration, or even exciting deviation – but the freak individuals with IQ 160 can be of paramount importance. Haldane and others have argued that the real human advances in the past were achieved by less than 1% of the population, by those who – as Mayr put it – were 'in the upper tail of the curve of human variation in inventiveness, imagination, perseverance, and ability to think clearly'. Once again there is the harsh alternative: 'a rather small downward shift in the mean value of the curve of human variability,' Mayr adds, 'might obliterate much of this highest class of potential achievers'. Is laissez-faire to prevail towards these changes or will man become a more purposeful master of his own fate?

Education is a tremendous force in our society. For good or for ill (and for ill read Ivan Illich) it dominates a great hunk of our lives. For those who will never be academic to any degree it consumes a decade; for those who require more qualifications it will eat into two decades, more or less, at the start of their lives. This is a tremendous proportion, ranging from a sixth to a third of everyone's lifetime. Education can boost a person's intelligence, by exercising his mind, by stretching its potentiality, while other environmental possibilities – better nutrition, better public health, better child care – can also help to raise an IQ. What they cannot do is to shift someone far below average to a position well on the other side of the midline. An 80 IQ will not become 140 however hard he tries, or society tries, or schools try. People are, in the major degree, born not made. As with congenital cripples of any kind they cannot, alas, shake off the bulk of their inheritance. The only possibility for large improvement is an attack upon the fundamentals, upon the genetic basis of each and every one of us. In every moment of conception we are, give or take a few environmental modifications later on, our unique selves. From then on the scene is set, inalienably.

Should intelligence continue to be our prime consideration for improvement of the species? Already over-promoted by education, and by no means the most important factor in any human relationship, should not other qualities be more actively pursued? To describe loved ones is to speak for an age without any recourse to their intellectuality, their academic record. They have patience, or humility, or love of life, or sensitivity, or charity, or fervour, or resilience, or curiosity, or generosity of spirit; but what of their IQ? Does it matter – at least within broad limits? Is it so crucial? School-teachers, busy casting sham pearls before real swine (to borrow Dorothy Parker's phrase with eternal gratitude), can be driven to tell a child it is no good merely because it fails to flourish in the classroom. Worse still, the child can believe this calumny. Worst of all, adults can be left with the impression that the mental ability most tailored to current educational requirements is the ultimate desirable quality, above selflessness, above creativity, above all.

Nevertheless, intelligence is welcome in a community; it would be hard to argue that stupidity is more important. More intelligence is more desirable; it would be yet harder to argue that greater quantities of stupidity are entirely satisfactory. In general terms, therefore, a more intelligent community would not be undesirable. How then, in more particular terms, could this be achieved? And who, for instance, would contribute the necessary germ-plasm? The image of that doctor, steadfast and erect in his confidence as to his suitability, comes quickly to mind. But who should take his place? And what is so desirable about medical students that they now contribute the bulk of the semen used for artificial insemination? They are correctly secretive about their contributions but are they, to cull a phrase from animal husbandry, of 'the right stock' to act in lieu of the infertile?

Muller, back in the 1930s, was in favour of Lenin and Darwin as possible founding fathers of a new generation. Others chose Abraham Lincoln and then had to reject him when geneticists suspected his long hands were the manifestation of a harmful recessive gene (Marfan's syndrome is characterized by a cardiac defect and lengthy fingers). Most heroes are most honoured after they are dead, their deaths having helped to enhance them; what then of their sperm? Other prophets are particularly unhonoured in their own countries; what of their sperm? What of Che Guevara, Albert Schweitzer, Jan Palach, or any of this century's holy men?

One answer would be to take donations from everyone likely to become desirable as a sire, creating a kind of *Who's Who* in sperm, a social semen register. As yet no sperm bank has acquired publicized offerings from any of the outstanding people of our time. Should, therefore, efforts be made to initiate such a collection? Or will people be unwilling to contemplate the germ of a Pasternak or a Casals for fear of the fraught circumstances such a conception might incur? They could well prefer, however un-enterprising, a more average offering, even some faceless and nameless medical student, even that steadfast practitioner.

The major argument against positive eugenics is that the successes of stock-breeding cannot be applied to man, either morally or scientifically. Peter Medawar, for example, argues this dictum in *The Hope of Progress*. On the first objection he rests his case on a single sentence: 'The moral-political answer is that no such regimen of genetic improvement (selective breeding) could be practised within the framework of a society that res-pects the rights of individuals.' That, he adds, 'should be sufficient'. It is more than sufficient if human reproduction does become akin to stock-breeding, but the choosing of semen for artificial insemination by the parents who will bear the conception, for instance, is not the husbandry of the cowshed. Nor is the counselling against genetic deterioration, as in a later chapter. Nor is the deliberate pairing of two individuals for the purpose of producing progeny of a definite kind, as might happen in the world of sport. There is no need for positive eugenics to become compul-sory in order for it to become effective. Also there need be no assumption that a moral-political veto by one generation will remain inviolate once their descendants begin to have their say.

On the scientific objection the case against is more complex. The stock-breeder's aim was always two-fold; he wished to have a good animal (good meat, good wool, good stamina) that would also, in its turn, repro-duce more good animals. In a general kind of a way this twin wish was fulfilled. Unfortunately, owing to the extreme genetic diversity that any animal inherits (relating to characteristics such as meat, wool and stamina), the dual fulfilment is only possible by populations and not by individuals. This breed of cows may produce more meat than that breed, but not every individual of this breed will be good; the skinniest may well have less meat than the meatiest of the other and skinnier breed. The animals, in short, do not breed true.

How then to make certain both that an animal is good and that its pro-geny are also consistently good? The answer, as every modern stock-breeder will be quick to relate, is to cross two relatively inbred breeds.

When carried out skilfully this procedure can result in all the offspring being better than both sets of parents in every desired degree – more and better meat, better hide, better disease resistance, and so on. Nevertheless, these most blessed progeny cannot themselves be used to create the next generation. If they were so employed they would produce a wide-ranging lot, with a random scattering of benefits, and with none of the reliable uniformity that they themselves possess. To get their kind again it is necessary to repeat the process, to cross-breed the two inbred strains as before. In other words, to borrow a conclusion from Medawar as well as the argument: 'Somatic and reproductive functions are separated: the eugenic end-product is reproducible at will, but does not reproduce.'

However, what goes on in a good breeding station, or in a laboratory, with back-crosses and hybrids and every other manifestation of hybrid vigour, is not what anyone is proposing for mankind. The advocated human procedures are less overwhelming, less absolute and less scientific. They are more akin to the Neolithic notion of mating best with best, or even best male with any old female. The race-horse studs are also, in the main, less scientific, mating good running stock with others that also have good running in their pedigree.

Even so, this method is enormously different to the free range style of human reproduction where anyone mates, more or less, with anyone else. The fact that we, again more or less, pair for life does not lessen the random manner of our original choice. No eugenicist has yet advocated pure-bred lines of humans, with cross-breeding only between selected strains. Instead, their advocations are humbler, suggesting that randomness can be harmful and much good could be achieved by sparing a thought for the possible progeny before embarking upon reproduction. A more scientific attitude might one day prevail but, at the moment, it is the first of Medawar's objections that holds most sway. Science, in a sense, can wait; the moral-political issues are already in our minds.

'In the name of morality and public security this practice should be condemned,' said the physiologist Magendie in 1847. What, one wonders, was the outrage? Steam trains? Public executions? Trade unions? It is hard to judge over such a gap in years. Actually he was ranting against the use of anaesthetics in surgery. In 1917 Margaret Sanger, pioneer of birth control and coiner of the phrase, was being gaoled for disseminating mere facts on contraception. Ten years later Julian Huxley was being personally rebuked by Lord Reith for mentioning the same subject on the British ether. By no means is any moral code immutable and a new set of tablets is being carved out all the time. Haldane, pioneer, leader of thought, revolutionary and protagonist of many causes, was 'against abortion at the mother's own request'. Nevertheless, America and Britain and the communist societies that were less alien to him now have laws permitting, more or less, abortion whenever a mother wishes it. Times change; needs change; and even eugenics is beginning to get a hearing that would have been unthinkable, say, thirty years ago.

In conclusion, therefore, to a chapter more of generalization than precise intent, it seems correct to give a few eugenicist leaders a final comment. With new techniques and facilities, said Muller, 'the way can be opened up for unlimited progress in the genetic constitution of man, to match and re-enforce his cultural progress and, reciprocally, to be re-enforced by it, in a perhaps never-ending succession'. Making the same kind of point, Ernst Mayr: 'Anyone familiar with the fantastically rapid evolution of the brain from the ape level to the *Homo sapiens* level is entitled to the fond hope that the apparent standstill in man's evolution is only a temporary phase and that a way can be found to initiate a trend towards an even greater future.'

So too, Julian Huxley: 'If, as I firmly believe, man's role is to do the best he can to manage the evolutionary process on this planet . . . fuller realization of genetic possibilities becomes a major motivation for man's efforts, and eugenics is revealed as one of the basic human sciences.' 'As long as no such (artificial) selection has been tried,' wrote Jean Rostand, 'nobody has the right to assign limits to human genius.' 'Britain,' said A. S. Parkes, on a more national level, 'will provide a progressively smaller proportion of the human race. If we cannot rely on quantity we must rely on quality. For this reason, if no other, serious consideration of the genetic improvement of the race should not be delayed indefinitely.'

The desires of today's crop of eugenicists will not be fulfilled by any generation still entrenched in the old style of procreation, the hit and miss that has been with us since time began. However, not everyone is so fortunate that he or she can take part in this particular lottery. Biology is today permitting their infertility to be less absolute. Sperm transfer can be performed for them or even, not so far ahead, fertilized egg (or embryo) transfer. Both insemination and counselling are obvious opportunities for mankind to have a greater hand in its own genetic future and in the advancement of the species. These two positive amendments to the archetypal way of life therefore form the material for the next section. A chapter on ethics will then, not unreasonably, follow hard upon its heels.

CHAPTER 16

Unnatural reproduction

Married infertility – Artificial pregnancy – AID –
Illegitimate products – New legislation – Single girls
and prisoners – Paternal perjury – Egg transfer

*'The importance of the great principle of Selection mainly lies
in the power of selecting scarcely appreciable differences . . .
which can be accumulated until the result is made manifest to
the eyes of every beholder'*

CHARLES DARWIN

In Britain there are about 450,000 marriages a year, involving of course 900,000 people. As about 12% of couples have an infertility problem 54,000 of these marriages will have trouble in acquiring children, namely 108,000 people. In the following year, with a similar crop of marriages, there will be another 108,000 people with the same difficulty. In theory, therefore, and after ten years of accumulated pairings, there would be over a million people regretting the wayward behaviour of their joint reproductive systems. In practice some of them manage to adopt babies, others provide foster homes and the numbers who remain painfully conscious of infertility must be less than that proportion of 12%. Nevertheless, the quantity of people involved is considerable even in one country. As Britain's population is about one-seventieth of the world's figure the total number must be roughly that degree more impressive.

Efforts are of course being made to lessen population growth, but tremendous work is also being applied, almost entirely in the richer countries, to reduce infertility. One couple's inability to have children can only be seen in its narrow context as that couple's personal disaster; it cannot be viewed, least of all by them or their practitioner, as a minor blessing for the planet. Therefore, everything will be done, and is consistently being done, to help them have the children they want. Many tens of thousands of offspring now owe their lives to modern methods of fertility promotion, and the figure will become many hundreds of thousands before so very long. Their parents profited from the new methods, and have thereby helped to populate the world still further. What is more intriguing is whether they are also, unwittingly, showing the way to a more eugenic control of future generations.

Where there exists an infertility problem the man is wholly responsible in 10–15% of the cases. Inadequate sperm numbers, faulty sperm, poor ejaculation or an inability of the sexual act itself can all lead to infertility. Nevertheless, and in many cases, an apparently infertile man is able to father children; his sperm can be centrifuged to concentrate their numbers, if this is necessary, or transplanted from his body into hers at the appropriate moment in her monthly cycle. If it is the woman who is infertile the probability is 50–60% that the error is caused by some lesion, such as a blockage of the Fallopian tubes or an adhesion outside either these tubes or the ovaries themselves. Sometimes the couple itself can be at fault. The two people may be both perfectly fertile with other partners but not, by the dictates of this perverse characteristic, with each other. They may even be psychologically disunited, ready for extramarital sex but not sex between themselves.

There is plainly every virtue in mimicking ordinary reproduction as far as possible if these people are to be helped, of using his sperm if they are good, of using her eggs or her uterus if they are good, and of using a couple's sperm *and* eggs if the fault lies elsewhere. In the past the insuperable problem of infertility was usually reconciled by adoption, but better contraception has lessened the supply (British adoption agencies placed 13,716 babies in 1967 and 8,417 four years later, a drop of 38%). Nevertheless, the kind of progress in reproductive studies that made the pill possible has also opened up the treatment of infertility. From the eugenic point of view the various methods either possible now or inevitable soon have differing potentialities. The list overleaf indicates which systems make use of the couple's own gametes and uterine environment and which need contributions from outside.

Of this list AIH is currently practised, although nothing like so frequently as AID. On the mother's side ETW will soon be performed successfully and there will be much talk from every side of the fence when that logical extension of AID does occur. Only when the dust has settled will the reproductivists move on to ETD and ETDD, not more difficult to achieve biologically but both likely to raise yet more furore when their turn comes.

Eugenic possibilities are incorporated within AID, ETD and, doubly so,

	Mother's genes	Mother's uterus	Father's genes
Normal conception	Yes	Yes	Yes
Adoption	No	No	No
AIH (Inseminating with husband's sperm)	Yes	Yes	Yes
AID (Inseminating with a donor's sperm)	Yes	Yes	No
ETW (Egg transfer using wife's own eggs)	Yes	Yes	Yes
ETD (Egg transfer using a donor's eggs)	No	Yes	Yes
ETDD (Egg transfer using a donor's eggs fertilized with a donor's sperm)	No	Yes	No

ETDD. The 10–15% of those 54,000 infertile British couples in each year's crop of marriages that can indict the male for failure mean that 5,400 to 8,100 couples have need either of AIH or AID. The other percentage – where the woman is mainly to blame – could probably be assisted by one of the three egg transfer methods. This is particularly likely in those 50–60% of cases where there is a blockage of the Fallopian tubes or a constriction of the ovaries. This could mean that over 20,000 women in each year's accumulation of marriages might be able to profit from the new techniques. Taking the men and women together the total is nearer 30,000, which suggests that about 60,000 new couples a year in Britain alone could transform their current infertility into much desired fertility. The numbers also indicate the considerable total, 70 times greater for the world at large, that could be susceptible to eugenic management.

The figures are guesswork. This is inevitable for future possibilities but, extraordinarily, no figures have been ever published totalling past AID endeavour. They are enshrouded, even if they exist, much like the abortion facts used to be. No AID register exists in any country. It is not as if we have had insufficient time to think about the problem: John Hunter, first in so many fields, was inseminating women artificially at the end of the eighteenth century. It is not as if the procedure was some intimate business concerned only with humans: in Britain and many other countries most cattle are conceived in this artificial manner, and at the end of the

1950s the Milk Marketing Board gave a banquet to celebrate the insemination of the ten-millionth cow in the United Kingdom.

The reason for the human secrecy, and the distinct lack of banquets about the one-hundred-thousandth human created in this fashion (the number guessed at in the United States alone up to 1957), is that the status of the act is extremely shaky. In Britain the operation is not illegal but the child is undoubtedly illegitimate. A case might be brought, although this has not yet happened in the courts, that the offending offspring was not liable to any inheritance. On the British birth certificate the father has to put his name in the appropriate column, but he is perpetrating a falsehood if he puts his own name when the actual spermatozoa came from another man. Nevertheless, although this problem is explained to him, he frequently does write his own name, this being the simplest course. He therefore commits perjury by this act.

In several American states and in France AID children are not illegitimate. (AIH children are always legitimate, despite the similar artificiality of their conception.) In Switzerland the practice is illegal. In Germany the AID child is legitimate until this is contested, perhaps by the child itself. He or she, in theory and in law, can claim maintenance, for example, from the natural father, the actual donor of the semen, provided he or she can run this man to earth. Again one should remember those men in South London who could not have been the fathers of 30% of their children, yet who are the putative fathers maintaining and loving the offspring in their charge. AID and the law would be performing a grave disservice if they combine to encourage husbands to check on the legitimate validity of the children they currently care for. If all sorts and manner of men are being run to earth, accused of sowing their oats wildly, and then being sued for maintenance, the result will surely be a loss of happiness in every quarter. So too with AID; if the law permits semen donors to become involved, the law will be doing wrong.

In 1960 a committee, set up under the chairmanship of the Earl of Feversham, reported to the relevant British ministers about the legal consequences of the growing procedure of artificial insemination. Extraordinarily, it recommended no changes in law relating either to legitimacy or birth registration. This committee also found, equally disconcertingly, that a large section of the medical profession and of the public disapproved of AID. The committee itself concluded that this new medical advance was *not* in the interests of society: therefore the practice should be strongly discouraged but not actually declared criminal. There were two committee dissenters to the idea of no legal change, and their recommendations are well worth quoting, if only because the rest of us have now largely caught up with their ahead-of-the-time opinion. They urged that:

1. The definition of legitimacy should be extended to include a child born as a result of AID to which the husband of the mother has consented.
2. For the purposes of registration of birth of such a child the husband should be deemed to be the father of the child.

Most current medical, judicial and ordinary opinion is also along these lines today. Therefore not many more decades shall probably pass before the necessary legislation is effected.

Eugenically we are probably less of one mind. Not everyone, for example, is in favour of medical students donating almost all the requisite semen. In a broadcast Bridgett Mason put the view in favour: they are readily available, they can be on call at unusual times, they are intelligent, they know what they are doing and they welcome the money. She expressed concern at using someone who actually volunteers to be a donor. On the other hand, and in a letter to the *BMJ*, John Slome argued that medical students were not the most satisfactory. 'The best donor is a man who has fathered two or three healthy children, which is the only way of knowing that his sperm are normal. Without exception, when I inform prospective AID recipients that only proved donor semen will be used, they are that much more reassured.'

It is easy to wonder whether their assurance would rise if they were handed a catalogue of curricula vitae belonging to potential donors. Perhaps they prefer the bran-tub of ignorance, if only because normal conception is much like that Christmas dip into the unknown to find presents good or bad. Or perhaps, as the AID parents are already experimenting with something new, they will flick through the list of biographies willingly. In both cases there are eugenic possibilities; someone has to choose the sperm. In the first instance the doctors choose, either from students or proven fathers; in the second the parents choose from the selection catalogued for them.

It is harder to appreciate why the public should have so resented the idea of AID in 1960 or even, to a lesser extent, today. Knowing the importance of children in family relationships, it must want infertile couples to have the chance, via donor masturbation and medical insertion, of conceiving children of their own. Can the very idea of reproductive interference be repugnant? Or, as other physical changes are largely impossible, is there envy that some people can better the ill fortune that their genes and their god gave them?

I also wonder what is wrong about advertising for donors. In 1972 the *Daily Telegraph* (of London) refused a doctor permission to place an advertisement for a suitable Hindu-Sikh. A year later the *Jewish Chronicle*, possibly even more surprisingly, made a similar refusal; a doctor was wishing to find a Jewish donor. Could it be that the editors of these journals were resenting the very idea of AID? If so, I wonder fearfully how they will treat the news of the first ETDD, the first human to start its life both within the laboratory and without the conceptual help of either of its parents.

The practitioners of AID are not totally in favour of the all-powerful role they have to play. If they give Jewish sperm to a Jewish girl because she – or her mother – insists upon it, should they also give Negro sperm to an Asian, or sperm from an athlete to a sporting enthusiast, merely because they too have requested it? On the other hand, although unwilling to accept every aspect of the god-like role, they are unwilling to yield very much. Yet why should the practitioner have any say in the kind of donor, apart from checking that he is devoid of defects? The future parents are far more concerned. An insemination is a few hours' occupation for the doctor, but a lifetime of involvement for the family.

In short, doctors' decisions are not always easy and should perhaps not be doctors' opinions at all. For example, there is the dilemma of AID and a single girl. She may not want a husband. She may not wish to make selfish seminal use of another man, yet she may want children desperately. Biological advance now permits a possible solution to her single problem, but it is not always granted. One doctor, stating during an interview that she had never granted AID to an unmarried girl, was plainly uncertain about her didactic attitude. Such a girl had once asked for the operation without first asking either her mother, father or brother with whom she lived. The doctor suggested she talked to them. Back came the girl with a mother wholly in favour of the idea. The doctor then asked if the family GP might not express an opinion. He did; he too was happy for the girl to have AID. Legally, it would be no crime for the insemination to take place, any more than it is a crime for a man to impregnate any single adult girl. Morally, it is hardly a crime, when the girl, her family and her family's doctor are all in favour of the scheme. Any subsequent child would be no more illegitimate than the AID offspring of a married couple although, and this is an important distinction, the child would have been born into a fatherless family.

It is not only doctors who have the power to refuse AID. The wife of a man serving an eighteen-year gaol sentence wrote to the British Home Secretary asking for artificial insemination via her husband. She knew she would be over forty when he came out again and that her time for having babies would also have been concluded. The written application for AID went through her doctor and back it came again – 'I regret to have to inform you that not sufficient grounds have been found for making an exception to the practice not to grant facilities for this purpose.' It is easy to suspect that a future generation, or even a future decade, will be astonished at the medieval thinking enmeshed within many of today's repressive actions.

So too with the concept of illegitimacy. An adopted baby, sharing none of its genes with those of its adopting parents, is deemed legitimate (in Britain) as soon as the parents have sufficiently confirmed their intent to the authorities and the actual mother has sufficiently relinquished her hold upon her offspring. With AID, as already mentioned, there is no such parallel, even though half the genes plus the mother's uterus are fulfilling their normal role. A man whose wife has been artificially inseminated should, strictly, put 'father unknown' in the birth certificate's 'Name of Father' column, thereby rendering the child illegitimate (defined as 'not in the legal position of those born in wedlock'). To acquire legitimacy it is then necessary for the couple to adopt the child; but what a business!

Most men opt for simplicity, giving their name to the registrar. Such men have already suffered the drawn-out ignominy of first learning and then fully understanding their infertility. They have had to accept the idea of semen from another male. They have witnessed the arrival of their wife's child and have possibly experienced much cradle-talk of family likenesses, of an uncle's ears and a grandmother's nose. The man who is going to love and maintain the child is then expected to go through the procedure of declaring the child illegitimate. How much simpler it must be for him to look the registrar straight in the eye and put his own name in the space provided.

The idea has already been touched upon of preserving semen from individuals of acknowledged greatness; but this procedure, quite apart from all the immediate difficulties of acquisition, would not be easy. New storage methods, new temperatures, new procedures for cooling the material are all the subject of current research; but, if they fail and if human sperm proves unco-operative over the years, each generation will have to resort to its own available selection of living donors. There will then be no question of reproductive reliance upon assured and venerated figures from the past.

There are other problems. An unproven AID sire might prove to be the possessor of unwelcome traits, such as harmful dominants or unwelcome quirks of personality. These might not be detectable for some years, and by then hundreds of his progeny could be in circulation. Another sire, rich with contributions to a given area, might be the cause of incest. Unwittingly, certain pairs of his offspring, half-brothers and half-sisters with more genes in common than first cousins, would elect to marry each other. It would be possible for one man to be responsible for 20,000 children a year, or so it has been calculated. If nameless and changing medical students are used there should be no problem on this score, but if the practice becomes commercial, and if certain idols of the day are suddenly and multitudinously desired as fathers, albeit at a distance and via insemination, the numbers of their descendants could abruptly outclass those of any harem-owning despot of the past. All the genetic perils of incest could then become most real.

It would appear that there are no adverse reactions to AID; the babies are no more abnormal than is normally the case. Nevertheless there are not the desirable numbers of follow-up inquiries. The parents, whose inseminated child has become firmly established as one of the family, do not entirely welcome probing questions over the years reminding them of the man-made artificiality they welcomed but wish to forget. If they come back for a second AID child, and a third, the possibility of checking on the earlier offspring is feasible. If they do not come back, for whatever reason, there is no such simplicity, and these cases might be considerably more revealing. However, it is a private matter, and it is respected for that fact. No one knows how many have been created and no one really knows how well they have fared. Here again it is easy to suspect that present attitudes will quickly change.

Egg transfer is a similar artificial procedure that will soon be involved in all the secrecy and legal confusion now surrounding sperm transfer. It will assuredly come to pass before society as a whole is prepared or competent to deal with it. First, the woman's own egg will be used and fertilized with her husband's own sperm before the union of both these gametes is inserted into the woman's own uterus. Secondly, and much later, there will be the same operation but with donor eggs and, if need be, donor sperm effecting the necessary fertilization.

Once again, the actual technique has an ancient history. In 1891 W. Heape, an English zoologist, extracted two four-cell embryos from an Angora rabbit and placed them in a Belgian hare. Shortly beforehand this recipient doe had been mated by a Belgian buck. Heape then waited for the results of his interference to be born and was delighted when the doe produced not just four Belgian hare offspring but two Angoras as well. Since then at least fourteen mammal species have experienced similar interference and countless surrogate mothers of different kinds have nurtured conceptions that were not their own. Rats, mice, ferrets, mink, sheep, goats, cows and deer have all become pregnant and given birth in this fashion.

Without any doubt the procedure will also work with mankind, or rather with the womankind whose reproductive systems prevent internal conception from taking place. Already human eggs have been fertilized by human sperm within the laboratory. And already, creating considerable excitement when the announcement was made in July 1974, three babies have allegedly been born following laboratory conceptions. Professor Douglas Bevis, British obstetrician, made the remark almost casually at a medical meeting that two children had been born in Europe and one in England following their conceptions performed *in vitro* under his care. Their ages then ranged from twelve months to eighteen months; therefore the conceptions had taken place in 1972. He also announced, which was equally startling, that he had discontinued this work because of the pressures being imposed upon him. These had come both from the medical community and from others – a newspaper had offered a very large sum for him to disclose all, such as the names of the parents involved. Unfortunately, Bevis has published nothing about his success.

The act is no more illegal than is AID but there are still complications. There is no question of adultery, as there is with AID, because – with ETW and ETD – the husband's own sperm are used. Nevertheless, just as a husband can dispute paternity because of AID, a wife could dispute maternity following ETW, claiming that the eggs were wrongly other than her own. Similarly, will she be committing a falsehood if she gives her name following ETD to the registrar of births as the mother of the child? Or does that column of the certificate correctly have to state the bizarre information 'mother unknown'?

Far more worrying is the prevalent suspicion that the act could lead on to cloning, or the production of carbon-copy babies. However, that also involves another technique, namely the manufacture of identical eggs by the substitution of their genetic material, and is properly another problem. There is quite sufficient difficulty with ETW on its own but, assuredly, a line of scientific research should not be stopped for ever just because a correct ethical ruling for its potential use has not as yet been determined. If some agreed code of conduct could be established, Professor Bevis and others could continue their research, the public would be less fearful and another kind of infertility would have been cured. In Britain alone 20,000 women, and therefore 20,000 men as well, could benefit from the new techniques.

Hereditary titles in the United Kingdom are not passed on to adopted children, whereas property is treated normally; there is no difference in its inheritance between adopted and natural offspring. As so much titular inheritance goes through the male line it would be logical, so long as people continue to be interested in such things, if the inheritance is equally valid whenever the man's sperm are involved, as in AIH, ETW and ETD. Legally, there are undoubted benefits if children arising as a result of these new methods could have precisely the same status as any adopted or natural infant. On this score time will, once again, probably correct the incorrectness of today.

ETD will not come immediately on the heels of ETW because it requires another woman to provide an egg. ETD provides less of a technical problem than ETW, mainly due to the potential mother having to submit to two interferences. With ETW the woman who will eventually be rewarded with the birth of a child has to supply the egg and has therefore to submit to the operation. Donating an egg is clearly on a different level to giving semen; two minutes of masculine masturbation in a back room is undeniably less demanding than the internal operation enabling a surgeon to extract the necessary eggs from an ovary. As operations go it is minor, but it requires general anaesthesia, a 24-hour stay in hospital, and an earlier treatment of gonadotrophin hormones. The laparoscopy is effected about four hours before ovulation would have occurred. If a man receives £25 for a masturbation, what price the donation of these pre-ovulating oocytes? Undoubtedly donors will be found – after all, kidney donors are making a far greater sacrifice – but an egg will never have the simplicity of a phial of spermatozoa.

Much of the ordinary as opposed to the medical resentment at these procedures may stem from the belief that they are ushering in some horrific world of biological experimentation. The journalist's readiness to refer to a test-tube whenever some reproductive advance is proposed may be the cause of much current alarm. There is fear of a brave new world, and an immediate belief that babies will be bottled like battery hens, that a ghastly inroad will have been made upon the accepted excellence of the procedure for conceiving and bringing forth a child.

Can this fear be valid for AID, merely because it permits infertile couples to achieve a measure of fertility? Can it be true of egg transfer, just because the actual fertilization takes place outside the mother's body? In both cases the mother not only carries the child but eventually gives birth to it. This is no battery procedure but is as natural as, in the circumstances, can be contrived. Perhaps some society other than our own, some alien breed of people far removed from any this planet has ever known, will one day be sufficiently callous to dispense with the uterus. Perhaps it will plan incubation chambers, and alphas and betas, and relish the virtues of Huxley's soma, and future women may welcome the withdrawal of their current reproductive role, but all such notions are entirely foreign to this particular century. Insemination is no more, and no less, than the substitution of good sperm and good eggs whenever deficiency of these blessings makes their replacement a total joy for those involved.

Far more likely are lesser manipulations. One advantage of *in vitro* (or laboratory) fertilizations, as opposed to those *in vivo*, is that the offspring's sex can be detected at that extremely modest and early stage in its development. An egg-transfer mother may desire, or indeed demand, that her pregnancy should be either the boy or the girl that she welcomes most assuredly. Also, just as mothers made use of wet-nurses in the past, it is not unimaginable that future mothers would pay helpful and needy ladies to become great with child on their behalf. Not every woman likes pregnancy, for it can be troublesome and exhausting and may, horror upon horrors, cause the permanent loss of a youthful figure. In fact some adoptions may have been preferred for this kind of reason and the substitute egg-transfer mothers would be fulfilling a similar role. However, these surrogates would actually be incubating the seed of those whose child it really is and whose child it will become.

To sum up, but from the eugenic point of view. Someone has to choose the sperm. Someone has to choose the eggs. Hundreds of thousands of couples are childless (almost a million adoption applications are made in the US annually with only a small fraction being satisfied) and hundreds of thousands of people will wish to experiment with the new techniques.

Gametes can be chosen with much more care than occurs naturally in the hurly-burly of mating and marriage. Sperm can be taken from donors who have shown no signs of genetic defects. Fertilized eggs can be inspected for certain types of congenital abnormality, such as mongolism, before being placed within the uterus. Sex-linked diseases can be better controlled, as the sex of the *in vitro* group of cells is detectable and the faulty sex can be eliminated. In almost every such advance in reproductive knowledge and technique there are eugenic possibilities. It is not a brave new world; it is merely the next world that is inevitably coming our way. It is we, after all, and the work being done now, that are creating it.

CHAPTER 17

Future counsel

Dominant – Recessive – X-linked – Genetic advice – Its
influence – Amniocentesis – Selective abortion – The PKU
problem – Parents *v.* embryos – Society *v.* individuals –
The trauma of information

'If we could first know where we are and whither we are tending, we
could better judge what to do and how to do it,' said Abraham Lincoln.
One wonders who has been most ignorant of the immediate future, the
people of his time or people now. Gregor Mendel was laying the foundations of genetics at the time of the American civil war; we today are
beginning to tinker with a score of possibilities raised by that Brno advance. Earlier chapters have touched on many of them; this one will
touch on several more.

Genetic counselling, for example, is now expanding rapidly and is
firmly based upon genetical knowledge. It has arisen because at least 4%
of live-born individuals suffer from some genetic or partly genetic defect.
Almost all – roughly 90% – of the inquiries received are from couples
who have had a disordered child and who fear a recurrence. This point,
therefore, must be resolved by the counsellor. He has to describe the
nature of the disorder, its genetic reason and the general scientific background to any counsel he can give. The parents may feel guilt at having
given birth to a faulty child – they often do – and this impression has to
be dispelled. Counselling is now more important than ever before if only
because congenital malformations are becoming relatively more important; they used to be responsible for 1 in 30 of all infant deaths (in Britain
in 1900) and now, with so many other causes of infant mortality successfully vanquished, kill 1 in 5.

The fundamental facts of inheritance provide basic guidelines for the
counsellor. Dominant inheritance is relatively simple; there is a 50%
chance that any child of an affected parent will be similarly affected.
Recessive inheritance is equally simple when both parents are known
to be carriers, this fact having been demonstrated either biochemically

258

or via the crucial proof of an affected child. In both instances the risk is 25% that any future child will be similar. Sex-linked inheritance is no less straightforward. If the gene is on the X chromosome, which has always been the case so far in sex-linked inheritance of disease, there is a 50% chance that each son will be affected and an equal likelihood that each daughter will be a carrier. There is, at first glance, a simplicity to counselling.

There is also, it must hastily be added, an extremely confusing side. There can be partial dominance associated with what is called penetrance that is of either a greater or lesser degree. Similarly, a gene may be inherited but it may also need an environmental trigger to set off its influence. Different social classes, for example, experiencing a variety of environmental triggers – diet, cold, disease – show a differing range of genetic abnormalities in consequence. There may be more than one gene affecting the disease, there may be mimic effects resembling the action of an inherited gene, and always there can be mutations, either occurring or even disappearing, thus confounding the predictions still further.

The difficulties are well demonstrated in the textbook *Genetic Counselling* (by Alan Carruth Stevenson and B. C. Clare Davison). For example:

'Migraine. This is such a common condition that familial cases are inevitable. Nevertheless there are more such cases than would be expected to occur by chance. Risks to first-degree relatives (such as children) are probably of the order of 1 in 10.'

'Marinesco-Sjögren syndrome. This syndrome of ataxia, cataracts and mental defect is almost invariably autosomal recessive, although a mild dominant type has been described.'

'Situs inversus (reversal of organs from their normal side). With or without dextrocardia (right-sided heart) this has been regarded as an autosomal recessive trait. However there are many sporadic cases and risks to sibs of unrelated parents are less than 1 in 4, probably about 1 in 15.'

By no means is there the simplicity of Mendel's garden peas, or even as much as might be wished by the counsellors, let alone the counselled, when genetic forecasts are called for.

Dominant inheritance is usually the most striking kind. Most of the unwelcome traits inherited in this fashion are not particularly harmful because, if they were, they would almost certainly have been eliminated by natural selection. Some can skip generations, making it harder for the predictors, and some can arise or disappear via mutations time and time again. One of the most unpleasant is Huntington's chorea, which gives rise to poor muscular control and mental deterioration during middle age. The victim has therefore probably passed on the faulty gene to some of his or her descendants before becoming aware of the affliction. Fifty per cent of the children can then expect to be victims, and they too may choose to have a family and therefore run the risk of transmitting the wretched gene once again. Other examples of dominant conditions are:

Achondroplasia	a form of dwarfism
Dystrophia myotonica	a muscle disorder
Marfan's syndrome	an eye and finger complaint
Osteogenesis imperfecta	a bone deformity (it afflicted Toulouse-Lautrec)
Thomsen's disease	a muscle disorder
Epiloia	an error of the central nervous system
Multiple exostoses	a disorder of the long and flat bones
Peutz-Jegher's syndrome	polyps on the intestine and patchy skin pigmentation

Recessive inheritance only manifests itself when a faulty gene is received from both parents. The most common and serious such condition in Britain and other western countries is cystic fibrosis of the pancreas. The victims usually die before their adult life has begun. In Europe about 1 in 2,000 children are affected by it; therefore 1 in 25 or so of the rest of us are carriers for it. As a curious related fact, referring back to the much earlier discussion about the need for resistance to killer diseases, the heterozygotes for cystic fibrosis suffer much less from tuberculosis than the rest of the population. When these single-gene carriers become detectable – the system of detection is not yet perfected – they can then be advised against mating with another such single-gene carrier. If they take this advice, as with the Tay-Sachs story in the introduction, they will be spared the harrowing odds of 3 to 1 against that their children will begin to manifest the disease; bad coughs, pneumonia, slow weight gain, poor food absorption, abnormal faeces. Other recessive diseases are:

Amaurotic family idiocies	many types exist
Sickle-cell anaemia	the malarial legacy
Phenylketonuria	PKU, the idiocy usually preventable on a phenylalanine-free diet

Galactosaemia	a metabolic disorder (cataracts, enlarged liver, mental deterioration) caused by infantile intolerance to lactose
Glycogen storage disorders	many kinds exist
Thalassaemia	the Mediterranean equivalent of sickle-cell anaemia

Sex-linked inheritance is much rarer. After all, there is only one pair of sex chromosomes and there are twenty-two pairs of the normal kind, the autosomes. Even so, over a hundred X-linked conditions are known in man – but not one that is Y-linked. Haemophilia is the most famous X example and it affects one male in every 10,000. There is also the Duchenne type of muscular dystrophy which, for identical reasons, affects males only. Its signs show either early (stumbling gait, difficulty with stairs) or much later (at 5–8 years). In the second case the disease's progress is likely to be more rapid, with few boys of this kind walking after they reach the age of twelve. Fortunately, about 80% of the – always female – carriers for this recessive disorder can be detected because of the abnormal enzyme levels in their blood. They can be advised of the results should they decide to have children and the odds will be the same as with haemophilia: 1 in 2 of their sons will suffer the dystrophy, 1 in 2 of their daughters will be carriers and will therefore be presented later on with the same hideous choice. Should this form of dystrophy occur in a family all the close female relations of the affected male, his sisters, aunts and nieces, will want to know if they are carriers. His mother will know that she is, unless the disease has been caused by a brand-new mutation and, if so, the sisters, aunts and nieces may be able to experience the relief of knowing they cannot be carriers. Other X-linked conditions are:

Christmas disease	a blood disorder, also known as haemophilia B
Glucose-6-phosphate dehydrogenase deficiency	a metabolic disorder, also known as G6PD
Nephrogenic diabetes insipidus	a form of diabetes

All such diseases form the bedrock of genetic counselling and recently the Department of Health and Social Security (in London) produced a small booklet, *Human Genetics*, which lists the thirty-five establishments in Great Britain where advice can readily be acquired. Plainly there is great need as the Oxford centre alone deals with 450 inquirers a year. The booklet also gives a rough guide to the recurrence risks for some of the commoner inherited conditions. For example:

Infantile pyloric stenosis (a blockage of the stomach exit)	1 in 20 (risk) for brothers or sons and 1 in 40 for sisters or daughters (of affected males). 1 in 5 for brothers or sons and 1 in 10 for sisters or daughters (of affected females)
Spina bifida cystica; Anencephaly (both errors of the central nervous system)	1 in 20 for siblings (of affected individuals)
Cleft lip (with or without cleft palate)	1 in 30 both for siblings and for children (of affected persons)
Congenital dislocation of the hip	1 in 40 for brothers and sons and 1 in 10 for sisters and daughters (of affected females). For relatives of affected males the risk is higher
Talipes equinovarus (club-foot)	1 in 50 for siblings
Down's syndrome (or trisomy-21 or mongolism)	About 1 in 100 for siblings independent of maternal age
Congenital malformation of the heart	When all types are taken together the risk for siblings is about 1 in 30
Diabetes (when onset under 30 years)	1 in 20 for siblings

The varying proportions of risk underline much of the difficulty inherent in genetic counselling and they also emphasize the problems for the couple being counselled. Is 1 in 20 so different in their comprehension from 1 in 30? Both the round number and the implicit vagueness in some of the risks add additional confusion, but genetics is rarely simple. For example, and making this point, the same booklet gives a list of common conditions 'with partial and complex inheritance'; some of them have already been encountered earlier in this book.

Congenital pyloric stenosis	Spina bifida cystica
Hare-lip (with or without cleft palate)	Anencephaly
Congenital dislocation of the hip	Diabetes
Congenital heart malformations	Schizophrenia
Manic-depressive psychosis	Ankylosing spondylitis

The counsellor asks about, or even looks at, as many relatives of each inquirer as he feels is necessary. People can be disinclined to mention other familial defects, or may not even know of their existence, but all such facts can be germane, particularly if others have suffered the same affliction as the individual who caused the inquirer's visit in the first place. For example, if parents have not had one but have had two off-spring with either spina bifida cystica or anencephaly, the recurrence risk for any future offspring goes up from 1 in 20 to 1 in 8. If a parent has a cleft lip and so does one of his (or her) children, the risk for further children rises from 1 in 30 to 1 in 10.

It would appear that people are much influenced by genetic counselling. In one Edinburgh study, of all those told they were at high risk (defined as greater than 1 in 10) of having an affected child, four-fifths decided to avoid a pregnancy and almost all of them did so. For those at medium risk (between 1 in 10 and 1 in 20) only a third decided to avoid pregnancy and all of them did so. Of the low risk people (with recurrence chances of less than 1 in 20) about two-fifths tried to avoid pregnancy and all but one did so successfully. Translating these proportions into figures, a total of fifty-seven subjects (including the high, medium and low) decided to avoid pregnancy in their marriages (fifty-two did so successfully), and another total of thirty-seven did plan pregnancies despite that forewarning about possibly producing affected children.

There were ten subjects at high risk who planned pregnancies despite the risk and they, as might be expected, had a variety of fortune. Five of them had a normal child, three had two normal children, one had one affected child, and one had two affected children. The genetic roulette had therefore created a total of eleven normal children and three affected offspring. These actual odds of 11 to 3, bearing in mind the prediction of worse than 10 to 1, form the sort of harsh reality underlying genetic counsel. There is no more certainty than there is with reds and blacks on the spinning wheel, and most of the couples were lucky (eight were and two were not in this series of ten), but there is an inevitability about the odds that does not lose its grip upon the population at risk. It is this point that the counsellor has to put over to the individual couple sitting before him. They see themselves as single people wishing to hazard one chance, perhaps two. He sees them much as a casino owner might see them; they are part of a group at risk. The individuals might win but the group will not. The dice are loaded, inexorably.

The hideous gamble linked with high risk, inheritance, and whether or not to go ahead with a pregnancy may diminish when amniocentesis becomes more widespread. Casinos give no chance of withdrawal once bets have been placed, but the procedure of extracting some amniotic fluid is more humane. For a large number of inborn errors, such as faulty chromosomes, the operation can reveal whether or not a gamble has been successful. It can also show, by indicating the sex of the developing foetus, whether or not a sex-linked and unwelcome character is also present: if the foetus is a male there is a 50% chance the sex-linked defect will be present; if female it will be absent but there is a 50% chance of it being a carrier. Even some metabolic disorders can be discovered via the amniotic fluid. In other words, if the gambler has been unfortunate, amniocentesis will be able to show in many cases that the risk has not paid off. There is then time for the foetus to be aborted and these luckless parents, if willing, can therefore try again. If the couple has been fortunate, the pregnancy can go ahead with infinitely less concern.

One recent series involving 331 patients (from Chicago) gives a good idea of this method and its possibilities. The amnion operation was performed in every case after the first three months of pregnancy had been concluded. The optimum time is sixteen weeks because it is difficult to obtain a sample of the amniotic fluid earlier when the target is so much smaller. It so happened that seventeen of the attempts failed to acquire any fluid at the first try and twenty-three others failed to realize a sufficient quantity; these forty patients were therefore subjected to a second and successful attempt. Of course there are risks in the procedure, both to the mother (bleeding, infection, blood group sensitization) and to the foetus (abortion, puncture, induced malformation), but there were no mishaps in this series.

The results and the reasons for the foetuses being studied were:

Reason	Pregnancies studied	Affected foetuses	Aborted	Delivered
Chromosomal	255	21	19	239
X-linked recessive	27	12 (males)	8	19
Metabolic	49	8	8	42
TOTAL	331	41	35	300

The number of foetuses both aborted and delivered make a greater total than the number of pregnancies studied because some pregnancies were of twins. The cause of the discrepancies between affected foetuses and aborted foetuses is, one assumes, because certain mothers did not wish for an abortion even though they had been advised of the facts. (It is possible, for example, to imagine a woman presuming she would want to have a malformed foetus aborted but then changing her mind when informed that she was carrying such a foetus.) Two children were born, for instance, with Down's syndrome and both these cases had been detected via their amniotic fluid. Every other child born at term was chromosomally normal.

It is therefore easy to wonder whether amniocentesis will grow faster in the future. The technique has improved immeasurably in recent years and it is effective in providing intrauterine diagnosis. The Chicago study was aimed at a high-risk group (119 of the mothers were over forty) and therefore its discoveries were of a high ratio. To find forty-one malformed foetuses within 331 pregnancies is abnormal but many malformations detectable via amniocentesis occur even among the ordinary pregnancies of low-risk people.

Certainly mongolism, for example, has a higher frequency among the births from older mothers but the majority of mongol offspring are born to younger mothers; the proportion is less but the numbers of younger mothers are so much greater that they in fact produce more mongols. The combined total, both from old and young, is one mongol for every 600 births, a formidable number for just one kind of chromosomal defect. If amniocentesis were made more widespread it could certainly pick out these and many other aberrant forms. The cost would be high but so is the cost, both in money and emotion, of keeping aberrants alive. About a third of the permanent inmates of British mental institutions are there due to mongolism, to that extra portion of chromosome No. 21. It would be preferable to inform the mother of the mental defective she is carrying, to arrange for an abortion, should she wish it, and to reduce the institutional intake of large numbers of otherwise inevitable clients. Britain alone creates over a thousand mongols every year, often much loved and generally very loving; but, if mothers had the choice, would most of them not opt for abortion? If so, they would then be ready for the normality of another child.

Patently, the parents should have most say. Does the mother want an amniocentesis and will she then abort the malformed foetus discovered within her? Medical opinion is generally firm. For example, it customarily recommends that amniocentesis should be offered to the patient only on the clear understanding that, if the foetus is found to be affected, the pregnancy will be terminated. The time is therefore somewhat distant before any women can go on her own volition for a comforting check up, perhaps on the sex of her future offspring, but it is likely to come, nonetheless. Abortion laws, like those in existence today, were unthinkable a dozen years ago. Amniocentesis laws, currently unthinkable, may also be with us before so very long. There will be similar detracting talk of 'convenience operations'; there will be inevitable mention of decadence and disgust; but worried parents will bring pressure to bear. Currently they accept the 2% of congenital malformations being born, but will they snatch at future opportunity to lessen this ratio, to relieve a possible burden upon themselves, to abort selectively? If strides are taken in this direction, for whatever reason, they will be eugenic strides. There will be genetic betterment, whatever the motive.

A degree of callousness seems to be an intrinsic partner of many reproductive advances. It is unfeeling, in a sense, to swallow the pill and decide when fertility is convenient. It is far more so, when contraception has failed or has not even been applied, to abort the error, despite its potent viability. Amniocentesis is either more or less humane, according to your point of view. It permits a preview, and then a choice. It is callous unless, of course, it is yet more callous to proceed with an unwanted child.

A similar dilemma, and considerable argument, concerns phenylketonuria, or PKU. The disease, which is a disorder of amino-acid metabolism, is inherited as an autosomal recessive – both parents must have passed on the faulty gene. It is not outstandingly common – about once in every 6,000 births in western countries – but it is currently the cause of an enormous screening programme. The reason is that, unlike so many other congenital disabilities, something can be done about PKU. If it is detected early enough, preferably within a few weeks of birth, and if the infant then receives a diet free of phenylalanine for several years, the victim stands a good chance of escaping the form of idiocy that would otherwise prevail. The testing method has been to examine urine, via nappies/ diapers, of as many babies as possible – Britain achieves a proportion of about 85% – in order to find the 1 in 6,000 who are the victims. New testing methods are on the way, making use of blood, but the determination is still strong to continue with the screening.

The question is whether it should be. The case in favour has been put well for the Manchester region. About 70,000 babies are tested there each year at a cost of £7,400. About eleven victims are encountered annually, and their very special diet, plus a portion of their dietitian's salary, adds up to £38,000 a year. These two totals, when divided by the number of affected infants, produce the answer of £4,100 a case. If these cases are untreated each victim is likely to spend at least thirty years in an institution. This cost, at £800 a year, would add up to £24,000 per victim. Therefore, on price alone, the screening programme is a financial blessing. Assuming these figures are applicable to the country (Britain) as a whole, and assuming also a similar proportion of PKU births throughout the land, the gross saving in detecting the annual number of about 130 victims is about £250,000.

The case against is equally forthright, seemingly more harsh, at first glance, and possibly yet more economical. Basically it points out that the PKU story of screening and prevention is not quite so clear-cut. Firstly, not every PKU victim can be saved; sometimes their brains are badly damaged by the time the special diet is implemented. Secondly, the average IQ of treated children is below normal. Thirdly, the special diet low in phenylalanine is exceptionally severe; most protein contains 5–10% phenylalanine and therefore special protein hydrolysates are prepared, to which are added appropriate fats, carbohydrates, minerals and vitamins. The infant is therefore eating artificially and without the traditional variety of food. His (or her) family has a definite burden in its midst and the diet has to be carefully maintained for the first five or six years of the child's life. A dietitian has to make regular checks of blood and urine because protein malnutrition, megaloblastic anaemia, loss of weight or even death can occur if supervision is imperfect. There is also a possibility that a treated PKU victim may, if female and eventually pregnant, harm the foetus she is then carrying because of the excessive phenylalanine levels still circulating in her maternal bloodstream. By no means, therefore, is the salvaging of a PKU person plain sailing.

The argument against concludes its case by recommending that parents and the state should forget about PKU, that the offspring should be encouraged to die and that another and healthier child should be conceived without delay. This practice would, one assumes, be heartache at the time but a new child will, if normal, be a firm blessing to erode most or all of that earlier unhappiness. From the state's point of view, and reverting to the balance sheet, it is plainly preferable neither to have PKU patients costing £4,300 for their treatment nor to have untreated victims costing £24,000 in their lifetime. Should the state be able to save on PKU, apparently callously, it can easily find another cause for the money and, possibly, a better cause into the bargain.

The PKU argument involves so many discordant aspects that it shows up many of the dilemmas. Is there a momentum about the struggle to save life that ought, on occasion, to be quelled? How much can a money argument be used to solve the problem of whether or not a particular medical approach should continue? (Of course this is happening all the time when hospitals allocate their funds, when grants are appropriated, but the scale of operation blurs the issues.) With PKU the choice is as stark as the frequent and saddening fund-raising for dialysis machines; without one a kidney victim will die, with one he will carry on.

Inevitably the PKU story, and every other dilemma mentioned in this book, lead on to a final discussion about ethics, and morals, and who is right and what is wrong. Morals are not unchanging. Each generation does shift its ground. Each generation has to do so because its problems are different. Consequently new ethical codes do have to be worked out; but never before, due to the accelerating pace of scientific and medical inquiry, have so many old and comforting precepts been susceptible to erosion. This subject therefore provides the material for the final chapter.

CHAPTER 18

Future ethics

A second explanation – Conflicting interests – Medical
opinion – Mental disease – Moral judgements – The
requirement for change

'*Today the world changes so quickly that in growing up we take
leave not just of youth but of the world we were young in*'

PETER MEDAWAR

'*The question is how to persuade humanity to acquiesce in its
own survival*'

BERTRAND RUSSELL

I do believe that we are giving more coherent thought these days than
ever before to the matter of human rights: we are less inclined to accept
dictates, religious or otherwise, from the past. However, in our efforts to
rearrange the situation, there is currently considerable illogicality, not to
say downright confusion, in what we say, think and do. We happily call
a friend an idiot unless he is one. We will call people cretins and morons,
and then be hideously ashamed of ourselves if we discover that such is
actually the case. We would never insult a mongol by calling him a mongol to his face, and yet we say it is insulting to the Mongoloid groupings
of mankind even to use that word. We say that money should be irrelevant
for any patient in need of expensive treatment, and yet our entire society
hinges upon it; perhaps the patient has only become a patient for a lack
of money in the past. We may even resent that PKU argument of the last
chapter because it was founded upon the rival costs of two possible procedures, but do we really consider that a life should be preserved at any
cost? And do we really think that a life already seventy years old is as
worthy of expenditure as one at the beginning of its span?

We are full of euphemism. People are unfortunate, backward, sick,
retarded; the true nature of the complaint is often as hidden as the victims
themselves. We build better homes for mental patients, and no longer
have the village idiot wild and feckless in each community, but we can

then conveniently forget about our mental patients. We take children to coal mines and steel works but never show them a mental home. We spend £100 million a year in England and Wales on running services for the mentally handicapped, and yet the state spends £300,000 in the same period for research into the causes, prevention and management of mental deficiency. We sweep everything except money under the carpet.

Undoubtedly, despite all that, there is also more action these days than ever before concerning human rights. The swing towards legal abortion is a prime example. Nevertheless there is, most independently, extreme reaction. We lavish intense care upon our infants the moment they are born, but do not even expend thought upon the manner of their conception. There is a hidebound quality in our attitude that the genetic lottery of who mates with whom, and of what manner of child is conceived, should be preserved at all costs; but we know that these costs can include great misery, considerable hardship, incarcerated lives and much expenditure of mere money. If a third of our mental institution inmates are mongols is it not strange that we do nothing to lessen that intake? And is it not stranger still that we are reluctant to listen even to the arguments in favour of reducing that supply? The world is fuller of facts than ever before and yet, for many of them most intimately concerned with human difficulty, we do not want to know.

My purpose in this book was to seduce a few more of us into thinking about the problem. The jolly tales about Pierino da Vinci and the odd king Friedrich, the fascination we have for breeding creatures such as dogs, the curiosity we possess about our own lineages, both past and yet to come, were all bait. They were all intended to draw us nearer to some of the awkward realities encased within genetics. I do not believe we can disregard them. I do not believe, once knowledge exists, that it can be set aside. Ignorance of the law, they say, is no excuse for default. Similarly, ignorance of crucial facts concerning us cannot be an acceptable premise for current practice. I have tried to seduce, and to give as many facts as my manner of writing would permit, and now the book is at its close.

'There is my truth; now tell me yours,' said Nietzsche. In short, what do you *now* think about the deliberate withdrawal of surgical aid from that mongol whose murder was mentioned in the introduction? What do you think you would do if you discovered that both your potential mate and yourself were carriers of Tay-Sachs? And what about all the other questions that pepper the previous pages? Their answers may not affect you personally, not yet; but the society you live in is going to have to deal with them, sooner or later, and probably sooner than most of us wish to think. Did you formerly consider that no child should ever be killed at birth, however gross its deformities? Do you still feel quite so sure? And do you feel, as we must all feel at times, that there is a slippery slope ahead of us if we make more changes? To kill any child at birth, or to let it die, or to refrain from surgery to prolong its hopeless existence, may be reluctantly commendable; but what of the next stage? How hopeless does that existence have to be? Is mental inadequacy the only criterion we wish to use? What of paralysis? What of physical deformity, such as half a face? Or three-quarters of a face? Or merely considerable ugliness? Or some minor unwantedness? At this point it is easy to throw up hands in horror simply at the mention of such gross absurdity, and to recoil further as one's own brain extrapolates other examples, such as murder for being the wrong sex, or colour, or degree of handsomeness.

However, it is plainly wrong to refuse some change merely because it will be hard to decide who or what should be affected by that change. Birth, for example, is so convenient; it enables us to say that pre-birth offspring can be legally aborted while post-birth offspring are legally protected. A man will be paid for performing an abortion, but a man may be gaoled for life if he kills a neonate. Is such convenience thinking still acceptable? Nature is adept at killing off poor products of conception – the malformed have, in general, a short life expectancy – but the simplicity of leaving as much as possible to nature, the judgements, the acts of mercy, the execution of the very sick, may no longer suffice as the only yardstick for our times. We are not cave-dwellers any more, haunted by fears and guided by magic and belief. We are a scientific people with access to facts. We cannot deny them. We have to yield to our knowledge of them. We cannot pretend, or should not pretend, that the old way of life is still with us. And, by the same token, we cannot or should not pretend that the old standards are somehow inalienable in the midst of all this change.

'I predict that in a hundred years' time,' Anne McLaren has written, 'we shall look back at the millions of congenitally malformed and mentally retarded children born in the twentieth century with the same pity and horror that we today view the high child mortality accepted as commonplace by our forbears.' The pity is felt for the parents who had to watch their children die. There was nothing else for them to do. Today we can control and assist and divert, and often we do nothing of the kind. Will therefore posterity look back with more pity and horror at our time, when we have so many capabilities, than at the earlier age when people were so powerless?

Of course there are reasons and explanations for our current behaviour. There are conflicting priorities, such as the interests of society today and of tomorrow. To bear and then to love a malformed baby can be commendable; to bear, to love and later on to hand that child over to the state, its burden having proved impossible, is of less appeal. A good many defectives live long after their parents have died, and other people therefore have to care for them. Whether state or family, these others did not make the choice eventually imposed upon them.

There are also the interests of the mother, the father and the embryo; these too do not always coalesce. A mother, having both carried and given birth to a child, may feel a greater wish to retain this product than the father who, although he welcomed the pregnancy, may be less welcoming and more rational when he sees what has been produced. And what are the interests of the embryo? Is it to be born alive at any cost? Foetuses who are permitted to live to term can, when living as independent children (and with someone to fight for them), sue for some alleged malpractice during their confined months in the uterus. But can or should they sue if their parents decide to disregard genetic counsel, to go ahead with a pregnancy, to hope that fortune and good genes will be on their side, and to know on the day of their offspring's birth that misfortune and a malformation are its inheritance? Some children have been known to sue (but not as yet successfully). The alternative to their deformed life would have been absence of life, and it is apparently hard for a court to declare that no life is better than some life. On the other hand, it is sometimes easier for the rest of us to make that point: we witness a great and genetic unhappiness and wish in consequence and for the sake of all concerned that it had never been born.

Often that offspring was conceived because the people concerned just did not know the true state of their personal situation. So how could they have acted otherwise? Barry Commoner, a compassionate Cassandra, is a great believer in supplying people with facts on which to base their opinions. The main job of the intellectual, he says, is to give sufficient information to permit others to be rational. 'We have no shortage of ideologies but there is a lack of factual grist for these various ideological mills.' With a subject as intimate and as personal as child-rearing, and with so much tradition inhibiting clear thinking about the new methods and the tremendous implications of biology, the facts ought to be given, he believes, both louder and clearer than they are at present.

Medical opinion is, on the whole, against passing on information that could be disturbing. Responsibility for saving, or for not saving, the life of a badly malformed baby is too heavy for the parents to bear, according to Professor J. P. M. Tizard. It should, he felt, be left to the conscience of individual doctors because this distressing aspect of medical ethics is better not discussed in public. On the same topic, and in the booklet *Professional Standards* (prepared in 1972 by a special panel appointed by the British Medical Association), it is concluded that 'the decision as to what is best in the interests of the patients must be left to the clinical judgement of the physicians and surgeons'. On AID, for example, and on another page of the same booklet, it is stated that 'a minority of the members of the panel could not accept this as an ethical procedure; other members thought it acceptable in certain circumstances, with the husband's consent as at present'.

Many of the rest of us can be amazed at such professional conceit. It does not hold out much AID hope for the single girl who wants a baby but not a man, or for the married couple who would merely prefer to use the sperm of another, or for the widow who has her husband's genes in cold storage, or for the woman who wishes a secret AID, being unwilling to inflict the knowledge of his sexual incompetence on to her husband. On malformation the same panel warns that 'the doctor must strike a balance between the trauma of giving information, and the trauma of allowing parents to have an abnormal child'. The decision can never be easy, but relatives can know a lot more than any doctor about the people with whom they live. Often they could better judge, provided the information came their way, how best to deal with it.

An irritation is that the medical profession is weak on ethical and moral instruction. Young doctors and nurses, on completion of their training, feel they have received little help in this regard (writes Professor A. S. Duncan in *Moral Dilemmas in Medicine*). Nevertheless, doctors and nurses will be expected to give a lead even though 'curricula have become so overloaded with the very scientific and technical matters which lead to the dilemmas that little time is left for thought as to the dilemmas themselves'. The rest of us worsen the situation by being yet weaker on moral and ethical instruction and by knowing even less about the scientific and technical matters. Our prevalent affection for chatting things over with the family GP, or with some earnest, tall-browed specialist, should not blind us to the fact that the dilemmas are ours, not theirs. The final decision may be theirs, and they may be extremely influential in their counsel, but the hazard of producing a genetically malformed child or of rearing such a child or of *being* such a child is in our court. The medical authorities handle the problem, but we live with it. Even if we eventually hand over our offspring to the state for permanent safe-keeping the grief stays with us.

A good example of all these points is mental disease. Or rather, following our happy tradition for disregarding the basic issue, and making use of the name in common usage, it is mental health that is the good example. Without doubt faulty genes are responsible for much faulty mental behaviour, and yet most of us keep the idea of genetics in quite a different compartment in our brains from the facts we know about mental health; but the two should be more closely allied. For example, genetic predisposition is a necessary (if not a sufficient) condition for the emergence of the schizophrenias and the manic depressive disorders. Also, if one parent is schizophrenic, the risk is 10% that the same disease will affect an offspring. Worse still, and aggravating the situation further, this risk continues throughout life; most schizophrenics do not show the disease to an appreciable extent until late adolescence.

Bearing this example in mind, and appreciating further the role of genetics in mental disease, some general points about this form of human breakdown not only have a disturbing flavour to them but this taste can be yet more worrying when one realizes how little the science of genetics is involved with the prevention of mental disorder. Read these sweeping statements, therefore, but think about genes:

At least 30% of all illness coming to the family doctor's attention is primarily mental, another 20% is secondarily so;

about 10–30 per thousand of any population are mentally retarded;

at least one in a hundred of any population is permanently suffering from a severe mental disorder, and at least 10 in a hundred will be so affected sometime during their lives;

one death in every 100 is by suicide (or a thousand people a day);

the cost of mental care is a high proportion of the total cost of health care (in the US it is 50%);

the risk of mental disorder increases each year in the elderly (people are now living longer, therefore mental disorder will increase commensurately);

no country has enough staff, according to the World Health Organization, to recognize and deal with more than a few of those needing mental care (some European countries have twelve psychiatrists per 100,000 population, some have one per 100,000, while most developing countries have only one per several million people);

more than one million patients are in European mental hospitals;

of the 60,000 men, women and children shut away in hospitals for the mentally subnormal in England and Wales, no more than 15,000 of them (according to the Office of Health Economics) suffer a degree of handicap so severe that they need 24-hour supervision and, of this number, just a few hundred need the kind of specialized medical care that can only be provided in a hospital.

If genetics can help to lessen the degree of any of these facts, so much the better for genetics. There is no continuing reason why the common clinical distaste for genetics should lead, as the psychiatrist Leon Eisenberg phrased it, to therapeutic pessimism. On the contrary, whatever it is that triggers psychosis and what might prevent that trigger being effective can best be discovered by the initial identification of people at risk. Genetics, via its basic laws and family studies, could provide that information. If it does so to any appreciable degree there will be every opportunity for lessening the fearful toll of misfortune behind every one of those general facts. Perhaps it is only a catalyst that is necessary to enable us to equate genetics more fully with the fortunes and mishaps that come our way. Or, as Donald Gould put it in *New Scientist*: 'In the shamefully neglected field of the care of the mentally handicapped, we may have to wait some such unhappy stimulus as the birth of an idiot grandchild to a prime minister.' After all, people's behaviour and the allocation of funds do not always (if ever) follow logical paths. It must have been highly relevant to poliomyelitis research, to the March of Dimes, and to our eventual conquest of the disease that its most famous victim was also the President of the United States, Franklin D. Roosevelt.

'Death has a function because you can't teach the old,' said H. G. Miller, Newcastle University's vice-chancellor, at a conference on the future of man. There do have to be new rules for new worlds. Medawar is also right that the world of our youth is as mortal as we are; the old tablets from the mountain do not last; the old standards have to change. We may like to think so, and think so increasingly with our advancing years, but there is nothing immutable about morals. They are 'that which is customary or generally accepted', and ethics is moral philosophy, or a concern for the nature of moral judgement. Neither ethics nor morals is therefore likely to be constant, either from society to society, or century to century, or even from the beginning to the end of our own humble span.

Conservationists have been accused of wanting the world to remain much as it was during their adolescence or early youth. It is said they do not wish the environment to return too far into the past, when predators were powerful, when undrained bogs were commonplace, or when woods were frightening and dark and often dangerous. They wish a compromise between now and then, and what better time existed than the dawn of their own existence? So too with many of the dilemmas encountered in all the earlier chapters. Do we really relish a time, as in the eighteenth century, when the British thought it wrong to number the people and take a census? Do we relish the nineteenth century when anaesthesia was condemned, or even the twentieth when doctors have been gaoled for giving abortions to very young girls who had been raped?

According to Malcolm Muggeridge, already quoted for his active reaction: 'From contraception to abortion to euthanasia is not, as I see it, humanitarianism broadening down from precedent to precedent, but a slippery Gadarene slide into another Dark Age.' Most people today accept the idea of contraception and are amazed at the recentness of much of the legislative change; many of this century's martyrs have merely been talking about it in public, or sending facts in the mail. A large proportion of people are in favour of abortion; they do not relish the act but realize its necessity and are astounded that the fight for legalization of pregnancy termination is still not won everywhere. Only a minority of people are currently in favour of euthanasia but most of us are happy to state that 'it was all for the best' or 'God was merciful' whenever some loved but very ailing individual suffers natural death. Do we really believe that we are on a slippery slope and that contraception, abortion and then euthanasia lead on to depths of darkness for our race? Or do we consider that we are doing the best we can in the light of changing circumstances, that contraception is a compromise between lust and overpopulation, that abortion is more welcome than the total unwelcomeness of unwantedness, and that euthanasia can sometimes be more humane than the inhumanity of nature?

Those who believe that this latter point is probably most accurate, that we do what we can to mitigate greater evils, will probably be more ready than most to accept that many medical morals are in need of overhaul. The difficulty is that we are each so intimately involved. As Eliot Slater said: 'If we didn't feel so strongly we might be able to think more clearly.' We do like to imagine, particularly in a welfare society, that money for medicine is unlimited, that dialysis for all in need is necessary (but Britain may be paying £40 million for its kidney patients every year in ten years' time), that every life should be prolonged so long as care and cash can lengthen it (but who needs more care and more cash than the very old?).

Quite plainly there is going to be more questioning in the future and already there is more than ever before. Here is Philip Rhodes, leading British gynaecologist, expressing general concern:

'I am very doubtful about efforts to prolong the life span, until we have solved the social problem of what the old can contribute to society except to eke out an existence. Research into the biology of ageing may be interesting, but what is to be done about its results? This raises the problem of screening for disease – for instance, by regular medical checks, mass radiography of the chest, well-woman clinics, and cervical smears. The value of all these, even to individuals, is doubtful and perhaps we need the courage to stop them and put the resources expended in them elsewhere.'

The task of removing some of these momentum expenditures is going to be formidable, and no less so than changing former opinion about the necessity for contraception and abortion.

Some change is infectious. If we decide to banish cervical smears, as being unworthy of the money spent upon them, we may be more in the mood for calling a halt to the nationwide testing for phenylketonuria, to the totally random manner of our procreation, to the production of so much genetic malformation. On the other hand, having had some change thrust upon us by the vigilant reformers, we may dig in our toes with resentment at the next similar proposal. The mounting restrictions on foetal research are surely one example. Both the US Congress and the California legislature suddenly (in 1973) forbade medical research on live aborted foetuses, despite the fact that foetuses have been exceptionally useful in the past, mainly in research dedicated to the better care of premature infants. He understood the motivation for the new law, said Dean Clayton Rich of Stanford University, 'but the bill is potentially more harmful than beneficial . . . Neonatal mortality remains disappointingly high. One of the reasons is our ignorance of foetal physiology'. Amply demonstrating the confusion of current thinking on this subject three Boston doctors were suspended in April 1974 by the Boston Board of Health and Hospitals for their work on legally aborted human foetuses. One week later, after the case had attracted national attention, the same board was prompt to reinstate the three men, who had been accused under an 1814 grave-robbing law.

Not only will old laws have to be brought up to date but we, in setting about this task and bringing pressure to bear, will have to bring ourselves up to date. We not only need facts but need to be able to argue rationally about this or that course of action. We will have to think of the needs of others, such as those who have produced malformed offspring or, yet more important, the offspring themselves. We will also have to think of ourselves – which will probably be far harder, of the need to extract facts pertaining to us personally, of the comprehension of those facts, of the actions to be taken which may need courage or cause resentment but which we, having given the problem coherent thought, consider to be right.

Human genetics has permitted a greater understanding of the old 'inborn errors of metabolism'. Genetic prevention, via counselling, is now wiser than ever before. Early detection, via amniocentesis, is now more feasible than ever in the past; more carriers of more genetic diseases can now be discovered before the carriers discover their possession the hard way – by the production of a malformed baby. In brief, without even embarking upon any deliberate eugenic programme, there are improved ways and means of reducing genetic error.

As so much illness and insanity are caused by wrongful genes, and as the national cost for these items is prodigious, any lessening of that account could pay dividends in many other quarters. Britain, for example, now spends £2,700 million a year on its National Health Service, approximately 5 % of its gross national product. Other countries spend a greater percentage, such as (in descending order of that percentage) Canada, United States, Sweden, Holland, West Germany and France. Every such service has to rob Peter to pay Paul, and the day has surely come for a more realistic assessment of the extent to which hopeless genetic defect has been robbing more hopeful possibilities. We looked at unwanted children, at back-street abortionists, at air embolism, at frightened mothers, and then – not before time – created new laws. We should therefore look just as hard at the equally suspect role of much genetic disease and realize that new guidelines, new standards and a new regard for suffering are now waiting urgently in the wings.

I think so. I thought so before beginning upon the research and the writing of this book. I think so even more now that it is concluded. To this end I have listed facts and have tried in those earlier chapters, not always with success, to suppress my views. To this end I have also had the hope that others, now more closely involved and enmeshed within the subject, will have a further foundation on which to build. There is my truth, once again. Now what is yours?

The glossaries

Two glossaries have been prepared, as explained in the introduction, in order to lessen the likelihood of thrombosis in the body of the text. The first is primarily about people, their dates, jobs and disciplines; whereas the second attempts to define and explain some unfamiliar words. Whenever the preceding pages already possess a description of an awkward term this is often thought to be sufficient; the information is not repeated in the glossary.

With people the barest descriptions have been given; the highest honours, the latest jobs, the most distinguishing details. Just enough fact has been listed to lend authority and relevance to statements from all those exploited in the text. The famous frequently achieve fewer lines because such people are well known; less need be said on their account.

With words only the meanings pertinent to this book have been given. In science there is a fearful duplication of meaning; terms much used in one field of endeavour have been robbed for others, time and time again. Think of nucleus, carrier, cell, base, negative, axis, and think of all their interpretations. With luck I have been as myopic as any other enthusiast for a subject, sticking firmly to the most relevant understanding and forgetting the rest.

The intention was always positive to make every description clear without there being need for recourse either to other entries or even nearby dictionaries. Occasionally this aim has been extremely fraught, not least with chemicals. Perhaps something is a phenol, and perhaps that degree of clarity seems inadequate. Phenol itself is also carbolic acid, and that is C_6H_5OH or even hydroxybenzene. It suddenly appears better to have stopped at phenol and to risk modest confusion, thereby offsetting the likelihood of extreme bewilderment.

The subsequent index is straightforward, save for an insistence that it should be legible. I personally and vehemently resent all indexes which mock their purported helpfulness by being printed in some minute lettering that only serves to antagonize. Similarly, there was no reason to detract from the glossaries by rendering them unreadable. Clearly they now follow.

Names

Adachi, Buntaro (1865–1945), Japanese geneticist, anatomist. Professor at Third Higher School, Kyoto, and Kyoto University; Dean of Medical Department; Director, Osaka Medical College.

Augustine, St (d. *c.* 604), Pope Gregory I's missionary to England; first Archbishop of Canterbury.

Avery, Oswald T. (1877–1955), Canadian bacteriologist, pathologist. Member of Rockefeller Institute for Medical Research, New York.

Ayer, Sir Alfred J. (b. 1910), British philosopher, author. Wykeham Professor of Logic, University of Oxford.

Bakewell, Robert (1725–95), British farmer; pioneer of livestock breeding experiments.

Barrett, Edward Barrett Moulton- (1785–1857), Despotic father of Elizabeth Barrett Browning, poet (1806–61).

Bateson, William (1861–1926), British biologist and founder of the science of genetics. Coined the word 'genetics'; founded the Genetical Society (1919); Director of John Innes Horticultural Institution, Merton, London; Editor of *Journal of Genetics*.

Beloff, Max (b. 1913), British historian, author, critic. Gladstone Professor of Government & Public Administration, University of Oxford.

Bernard, Claude (1813–78), French physiologist. Founder of experimental medicine; 'Préparateur' and later colleague of Magendie (q.v.); President of Académie des Sciences.

Bernhardi, Friedrich von (1849–1930), German cavalry general, militarist, author.

Besant, Annie, *née* Wood (1849–1933), Irish theosophist, traveller, reformer, writer. President of Theosophical Society; co-editor with Charles Bradlaugh (q.v.) of *National Reformer* under pseudonym Ajax; prosecuted in 1877 for republishing C. Knowlton's *Fruits of Philosophy*; winning their case they effectively achieved freedom of the press.

Bevis, Professor Douglas, British obstetrician and gynaecologist. Reader in Obstetrics and Gynaecology, University of Sheffield, England.

Binet, Alfred (1857–1911), French biologist, psychologist. Director of the Laboratory of Physiological Psychology, University of the Sorbonne, Paris.

Blumenbach, Johann Friedrich (1752–1840), German anthropologist. Extraordinary Professor of Medicine, Göttingen, from 1776; introduced race classification on a quantitative basis through his study of comparative skull measurements.

Bodmer, Walter F. (b. 1936), British geneticist. Professor of Genetics, University of Oxford; previously Professor of Genetics, Stanford University, California.

Bond, Charles John (1856–1939), British surgeon and sociologist. Helped to promote Sociological Society; was responsible for starting Voluntary Euthanasia Society; former Vice President of Eugenics Society.

Bourne, Aleck W. (b. 1886), British gynaecologist. Lately consulting gynaecologist, St Mary's Hospital, and Obstetrics Surgeon, Queen Charlotte's Hospital, London.

Bradlaugh, Charles MP (1833–91), English secularist, social and political reformer. Leader of Free Thought and Radical Movement; worked for limitation of population; co-editor with Annie Besant (q.v) of *National Reformer* under pseudonym Iconoclast.

Brazziel, William F., American educator. Professor of Higher Education, Connecticut University, Storrs.

Bruce, Hilda M., British biologist. Department of Experimental Medicine, Cambridge; lately Member of Scientific Staff, Medical Research Council, Mill Hill, London.

Burton, Sir Richard F. (1821–90), British explorer and writer.

Butler, Neville R. (b. 1920) British paediatrician. Professor of Child Health, University of Bristol, England; Director, Perinatal Mortality Survey.

Campanella, Tommaso (1568–1639), Italian philosopher, author.

Carlyle, Thomas (1795–1881), Scottish essayist, historian, philosopher.

Carter, Cedric O. (b. 1917), British geneticist. Director, Medical Research Council Clinical Genetics Unit, Institute of Child Health, London; President (1972–3) and past General Secretary of Eugenics Society; seven children.

Cavalli-Sforza L. L. (b. 1922), Italian geneticist, author. Head of Genetics Department, Stanford University, California.

Celler, Emanuel (b. 1888), US Congressman. Chairman of House Judiciary Committee (1924); responsible for liberalizing immigration by Celler Civil Rights Act (1965).

Chamberlain, Houston Stewart (1855–1927), British (naturalized German in 1916) political theorist, intellectual of Nazism, author. His *Foundations of the Nineteenth Century* was published (in German) in 1899.

Child, George P. (b. 1908), American physician, pharmacologist. Lately held appointments in US university and public health service.

Cole, Sonia (b. 1918), British writer. Associate of British Museum (Natural History).

Comfort, Alexander (b. 1920), British biologist, poet, novelist. Formerly

Director of Research in Gerontology, University College, University of London; now resident in United States.

Commoner, Barry (b. 1917), American biologist, educator. Chairman, Department of Botany, Washington University, St Louis; Director, Center for the Biology of Natural Systems.

Conklin, Edwin Grant (1863–1952), American biologist; early supporter of eugenics. Emeritus Professor of Biology, Princeton University, New Jersey.

Cooke, H.E. Cardinal Terence (b. 1921), Roman Catholic Archbishop of New York.

Coon, Carleton S. (b. 1904), American anthropologist. Research Curator of Anthropology, University Museum, Philadelphia, Pennsylvania; formerly Professor of Anthropology and Curator of Ethnology, University of Pennsylvania Museum.

Correns, Karl (1864–1933), German botanist, author. Director of Kaiser Wilhelm Institute of Biology, Berlin, Germany; then Professor of Philosophy, Berlin; rediscovered Mendel's law of inheritance (1900).

Crew, Francis Albert Eley FRS (1886–1973), British geneticist. Late Emeritus Professor of Animal Genetics, and Professor of Public Health and Social Medicine, University of Edinburgh, Scotland.

Crick, Francis H. C. FRS (b. 1916), British molecular biologist. Medical Research Council Laboratory of Molecular Biology, Cambridge; Nobel prize (1962 with James Watson, q.v.).

Crossman, Richard H. S. (1907–74), British Labour politician.

Crow, James F. (b. 1916), American educator, population geneticist. Lately Professor of Medical Genetics, now Acting Dean of School of Medicine, University of Wisconsin, Madison; President of American Society of Human Genetics (1963).

Curie, Marie, *née* Sklodowska in Poland (1867–1934), French physicist. Nobel prizes (1903 and 1911).

Dahlberg, Gunnar (1893–1956), Swedish human geneticist, author. Head of State Institute for Human Genetics and Race Biology, Uppsala University, Sweden; founded *Acta Genetica et Statistica Medica* in 1948 (now called *Human Heredity*).

Darlington, Cyril Dean FRS (b. 1903), British botanist, author. Emeritus Professor of Botany, University of Oxford; Keeper of Oxford Botanic Garden; President of Genetical Society (1943–6).

Darré, Richard Walther (1895–1953), German politician. Born in Argentina; emigrated via England to Germany; Minister of Agriculture; head of Race and Resettlement Office in Himmler's ss; dismissed (1942) when food supplies grew short; captured by Allies (1945); given seven-year sentence at Nuremberg trials, of which he served one year.

Darwin, Charles R. (1809–82), English naturalist, author. Expounder of evolution by natural selection.

Davenport, Charles B. (1866–1944), American biologist, embryologist, geneticist; one of leaders of eugenics movement. Founded Station for Experimental Evolution (1904) and Eugenics Record Office (1910); they merged in 1918 into the Department of Genetics of the Carnegie Institution, Washington.

David, Paul R. (b. 1907), American zoologist, human geneticist. Professor and Director of Medical Genetics, School of Medicine, University of Oklahoma, Norman.

Davison, B. C. Clare, Gynaecologist and obstetrician, Population Genetics Unit, Medical Research Council, Oxford.

Dean, J. Geoffrey K., British physician. Director, the Medico-Social Research Board of Ireland, Dublin; formerly Senior Physician, Eastern Cape Provincial Hospital, Port Elizabeth, South Africa.

Devereux, Robert, Earl of Essex (1566–1601), British soldier. Favourite of Queen Elizabeth I; beheaded for treason.

Dillingham, William P. (1843–1923), American lawyer; Governor of Vermont; Senator. Head of Commission investigating immigration (1907–10); author of Bill restricting immigration to US (1921).

Disraeli, Benjamin, Earl of Beaconsfield (1804–81), British prime minister; novelist.

Dobzhansky, Theodosius FRS (b. 1900), Russian (naturalized American) educator, zoologist, author. At Davis Campus, University of California.

Down, J. Langdon H. (1828–96), English physician. Held appointments at Earlswood Asylum, London Hospital and Normansfield Training Institute, Hampton Wick, England; first described mongolism (1866); author of *Mental affections of childhood and youth* (1887).

Drinan, Robert F. SJ (b. 1920), American Roman Catholic priest, educator, lawyer. Dean of Boston College Law School, USA.

Duhamel, Georges (1884–1966), French novelist, poet, playwright.

Duncan, Archibald S. (b. 1914), Scottish obstetrician and gynaecologist. Executive Dean of the Faculty of Medicine and Professor of Medical Education, University of Edinburgh.

Ehrlich, Paul R. (b. 1932), American biologist, educator, author. Professor and Director of Graduate Studies, Department of Biological Sciences, Stanford University, California.

Einstein, Albert (1879–1955), German (naturalized American) mathematical physicist. Propounded theory of relativity.

Eisenberg, Leon (b. 1922), American psychiatrist. Department of Psychiatry, Massachusetts General Hospital, Boston, Mass.

Emerson, Rollins A. (1873–1947), American geneticist. Professor of Plant Breeding, Cornell University, New York.

Engels, Friedrich (1820–95), German socialist, political philosopher, writer. Spent much time in England where he collaborated with, and edited, Karl Marx.

Ernle, Roland Edmund Prothero, Baron (1851–1937), British author, politician.

Etzioni, Amitai Werner (b. 1929), German-born sociologist, author. Professor and Chairman, Department of Sociology, Columbia University, New York.

Eysenck, Hans Jurgen (b. 1916), British psychologist, educator, author. Professor of Psychology, Institute of Psychiatry, University of London.

Fenner, Frank J. FRS (b. 1914), Australian biologist. Director and Professor of Microbiology, John Curtin School of Medical Research, Australian National University, Canberra.

Fermi, Enrico FRS (1901–54), Italian physicist. Emigrated to USA in 1938; researched at Columbia, Chicago and Los Alamos; set in operation first atomic pile; Nobel prize (1938).

Feversham, Charles William Slingsby Duncombe, Earl of (1906–63), British politician. Parliamentary Secretary, Ministry of Agriculture and Fisheries, London (1936–9).

Fischer, Eugen (1874–1967), German geneticist, biologist, anthropologist. Director of Kaiser Wilhelm Institute for Anthropology, Human Genetics and Eugenics, Berlin.

Fuller, John L. (b. 1910), American psychobiologist, author. Lately Associate Director, Jackson Laboratory, Bar Harbor, Maine.

Galton, Sir Francis (1822–1911), Versatile English genius, explorer, meteorologist, geneticist and founder of eugenics. Related to Charles Darwin.

Gardiner, Gerald Austin, Baron, QC (b. 1900), British lawyer. Chancellor of the Open University, Milton Keynes, Bucks., England.

Gates, R. Ruggles FRS (1882–1962), British biologist, botanist, geneticist, anthropologist. Professor of Botany, King's College, University of London; founder member, Genetical Society.

Gauss, Johann Friedrich (1777–1855), German mathematician, astronomer, physicist. Excelled in theory of numbers.

Ginsberg, Benson E. (b. 1918), American educator, geneticist. William Rainey Harper Professor of Biology and Associate Dean, University of Chicago, Illinois.

Goodhart, Charles Burford (b. 1918), British geneticist. Fellow and Director of Studies in Biological Sciences, Gonville and Caius College, Cambridge; founder member, Society for the Protection of Unborn Children.

Gordon, Cecil (d. 1960), British geneticist. Lecturer in Animal Genetics, University of Edinburgh.

Grant, Madison (1865–1937), American lawyer, naturalist, author. Promoter of eugenics movement, prominent racist and advocate of immigration restriction; conservationist; friend of Theodore Roosevelt (q.v.).

Gregory I, Pope (c. 540–604), promoter of Christian Church in western Europe; reformer of liturgy.

Grier, George (b. 1882), American radiologist. Senior Associate and Project Director of the Washington Center for Metropolitan Studies; President of Roentgen Ray Society (1934) and of Radium Society (1935).

Guttmacher, Alan (1898–1974), American physician. President, Planned Parenthood of America (1962); former Emeritus Professor, Obstetrics and Gynecology, Mt Sinai Medical School.

Hagen, Louis (b. 1916), German born; emigrated to Britain in 1936. Journalist, author and director of firm making children's films.

Haldane, Professor J. B. S. FRS (1892–1964), British biologist, statistician, author, radical and journalist. Late Head of Genetics and Biometry Laboratory, Orissa, India; founder member, Genetical Society.

Hammond, Sir John FRS (1889–1964), British animal breeder, physiologist. Superintendent, Animal Research Station, Cambridge; largely responsible for introduction of artificial insemination to British cattle-breeding industry.

Hardin, Garrett J. (b. 1915), American biologist, ecologist, author. Professor of Biology, University of California, Santa Barbara.

Harington, Sir John (1561–1612), English courtier, writer, wit. Invented water closet.

Harland, Sydney Cross FRS (b. 1891), British plant geneticist. Director, Plant Breeding Station, Hacienda Naña, Peru.

Hatcher, Geoffrey W., British consultant paediatrician, Sussex groups of hospitals, England. Worked on investigation of spina bifida children at Sheffield.

Heape, Walter FRS (1855–1929), English zoologist; originally businessman. Demonstrator in animal morphology; concentrated on research from 1893; partner in Heape and Grylle Rapid Cinema Machine (1915); founder member, Genetical Society.

Heath, Edward R. G. (b. 1916), British Conservative politician.

Hegel, Georg Wilhelm Friedrich (1770–1831), German philosopher and author, particularly of works on the human mind.

Herrick, James B. (1861–1954), American physician. Professor at Rush Medical College, University of Chicago, Illinois; discoverer of sickle-cell anaemia (1905).

Herrnstein, Richard J. (b. 1930), American educator, author. Professor and Chairman, Department of Psychology, Harvard University, Cambridge, Mass. (to 1971).

Hogben, Lancelot FRS (b. 1895), British biologist, statistical theorist, author. Emeritus Honorary Senior Fellow in Linguistics, Birmingham University.

Höhne, Heinz (b. 1926), German author. Joined *Der Spiegel,* where now Serials Editor; wrote *The Order of the Death's Head,* the story of Hitler's SS.

Huff, Darrell (b. 1913), American magazine editor, then freelance writer. Graduate of State University of Iowa.

Hulse, Frederick S. (b. 1906), American anthropologist. Professor of Anthropology, Arizona State University, Tempe.

Hunter, John FRS (1728–93), British surgeon, teacher, lecturer. House surgeon, St George's Hospital, London; made many original experiments in anatomy, comparative biology, pathology, physiology.

Hunter, Richard A., British psychiatrist, author (son of Ida Macalpine, q.v.). Lecturer, Institute of Neurology, London.

Huskins, C. Leonard (1897–1953), Canadian plant cytogeneticist. Late Guggenheim Fellow, Department of Zoology, Columbia University, New York.

Hutt, Corinne, British research psychologist. Senior Research Fellow, Department of Psychology, University of Keele, England.

Huxley, Aldous L. (1894–1963), English writer; brother of Julian (q.v.); grandson of Thomas (q.v.). Author of *Brave New World* (1932) describing a science-controlled future.

Huxley, Sir Julian S. FRS (b. 1887), English biologist, author. First director general of UNESCO; founder member, Genetical Society.

Huxley, Thomas Henry FRS (1825–95), English biologist; vigorous promoter of Darwin's theory of evolution.

Illich, Ivan (b. 1907), Austrian author. Founder and Head of Center for Intercultural Documentation, Cuernavaca, Mexico.

Inge, William Ralph (1860–1954), British priest. Dean of St Paul's, London; known as 'the Gloomy Dean' for the pessimistic tone of his writings and preaching.

Jefferson, Thomas (1743–1826), US Republican President (1801–9). Author of Declaration of Independence (1776).

Jenner, Edward (1749–1823), English doctor. Discoverer of vaccination against smallpox.

Jensen, Arthur R. (b. 1923), American educator, psychologist. Professor of Educational Psychology, University of California, Berkeley.

Johnson, Dr Samuel (1709–84), English writer and lexicographer.

Jordan, David Starr (1851–1931), American educator, naturalist, philosopher, author; one of leaders of American eugenics movement. Professor of Zoology, then President, Indiana University; later President, then Chancellor, Stanford University, California.

Joseph, Sir Keith S. (b. 1918), British Conservative politician.

Kac, Mark (b. 1914), Polish (naturalized American) mathematician, educator. Emigrated to USA in 1938; Professor of Mathematics, Rockefeller Institute, New York, from 1961.

Kahn, Herman (b. 1922), American defence analyst, author; 'Professional Prophet'. Co-founder and director of Hudson Institute, White Plains, New York, from 1961; pioneer of science of 'futurology'.

Karman, Harvey, American psychologist, Los Angeles. Inventor of cannula technique of abortion.

Kelvin, William Thomson, Baron (1824–1907), British physicist. Worked in thermodynamics, electricity and magnetism, proposed the Kelvin temperature scale, and invented the mirror galvanometer and quadrant electrometer.

Ketchel, Melvin M. (b. 1922), American physician. Director of Oak Ridge Population Research Institute, Tennessee.

Knowles, John H. (b. 1926), American physician and medical administrator. Head of Rockefeller Foundation, New York.

Koestler, Arthur (b. 1905), Hungarian (naturalized British) author and journalist. Member of *Graf Zeppelin* polar expedition (1931); imprisoned by Franco during Spanish civil war; served in French Foreign Legion and British Army during World War II.

Kornberg, Arthur FRS (b. 1918), American biochemist. Professor and Head of Department of Biochemistry, Stanford University, California; Nobel prize (1959).

Lacy, Dennis A. (b. 1927), British zoologist. Professor of Department of Zoology and Comparative Anatomy, St Bartholomew's Hospital Medical College, London.

Landauer, Walter (b. 1896), German (naturalized American) geneticist. Studied genetics and zoology at Heidelberg University in early 1920s; Honorary Research Fellow, Department of Animal Genetics, University College, London.

Landsteiner, Karl (1868–1943), Austrian pathologist. Left Europe in 1923 to work at Rockefeller Institute for Medical Research, New York; Nobel prize (1930); first to describe blood groups (1900).

Laplace, Pierre Simon, Marquis de (1749–1827), French astronomer, mathematician. Worked on lunar theory, proving long-term stability of the Solar System.

Laughlin, Harry H. (1880–1943), American eugenicist, trained as geneticist. Leading promoter of immigration restrictions and of sterilization laws; editor, *Eugenical News*; superintendent, Eugenics Record Office, Cold Harbor, Long Island, New York (he retired in 1940 when it was closed).

Laughlin, William S. (b. 1919), American educator. Professor of Anthropology and Science, University of Connecticut, Storrs; Chairman of Human Biology Committee, Human Biology Program (1966–9).

Leakey, Louis S. B. (1903–72), Kenyan anthropologist, author. Late Director of the National Centre of Prehistory and Palaeontology, and Curator of Coryndon Memorial Museum, Nairobi, Kenya.

Lidbetter, E. J. (1878–1962), British eugenicist, Poor Law Officer, author of *Heredity and the Social Problem Group* (1933).

Lippmann, Gabriel FRS (1845–1921), French physicist. Nobel prize (1908) for invention of photographic reproduction of colour.

Loomis, William Farnsworth (b. 1914), American biochemist, educator. Professor of Biochemistry, Brandeis University, Waltham, Massachusetts.

Ludmerer, Kenneth M., American author. Educated at The Johns Hopkins University, Baltimore, Maryland.

Macalpine, Ida (1899–1974), British psychiatrist, medical historian. Lately psychiatrist, St Bartholomew's Hospital, London.

McBride, William G. (b. 1927), Australian physician. Consultant Obstetrician and Gynaecologist, The Women's Hospital, Sydney.

McLaren, Anne D. L. (b. 1927), British geneticist. Director of Medical Research Council Mammalian Development Unit, University College, London.

Magendie, François (1783–1855), French physiologist, doctor; 'father of experimental pharmacology'. President of Académie des Sciences (1837); later President of new Advisory Committee on Public Hygiene.

Malthus, Thomas Robert FRS (1766–1834), British curate, economist, author. Professor of Political Economy, Haileybury College, England, from 1805; his *Principles of Political Economy* was published in 1820.

Marr, Wilhelm (1819–1904), German journalist. In 1873 published *The Victory of Judaism over Teutonism* (12 editions in 6 years); introduced term 'anti-Semitism' in Germany.

Mason, Bridgett A., British doctor. Clinical assistant, Obstetrics, Gynaecology, Fertility, London Hospitals Group.

Matsunaga, Ei, Japanese geneticist. Department of Human Genetics, National Institute of Genetics, Yata, Japan.

Maxwell, James Clerk (1831–79), Scottish physicist. First Professor of Experimental Physics, University of Cambridge (1871); principally researched in electricity and magnetism.

Maynard Smith, John (b. 1920), British geneticist; graduated in engineering, then zoology; researched into genetics, physiology and *Drosophila*. Dean and Professor of Biological Sciences, University of Sussex, Brighton, England.

Mayr, Ernst W. (b. 1904), German (naturalized American) educator. Professor and Director, Museum of Comparative Zoology, Harvard University,

Cambridge, Mass.; previously assistant curator of Zoology Museum, University of Berlin.

Mead, Margaret (b. 1901), American anthropologist, lecturer on child psychology and education. Adjunct Professor of Anthropology, Columbia University, New York.

Medawar, Sir Peter B., FRS (b. 1915), British biologist, author. Former Director of National Institute for Medical Research, Mill Hill, London; Nobel prize (1960).

Mendel, Gregor Johann (1822–84), Austrian Augustinian monk, botanist, beekeeper, meteorologist. Failed to qualify as teacher due to examination amnesia; Abbot of Brno monastery from 1868; first to propound laws of inheritance (in 1865).

Middeldorf, Ulrich A. (b. 1901), German art educator, historian. Adjunct professor, New York University; Director, Instituto Longhi, Florence.

Miescher, Friedrich (1844–95), Swiss biochemist. In 1869 isolated a substance containing both nitrogen and phosphorus from remnants of cells in pus; it was first named nuclein, later nucleic acid.

Miller, Henry George (b. 1913), British physician, neurologist. Previously Professor of Neurology at Newcastle University.

Mitford, Nancy (1904–73), British author.

Moivre, Abraham de FRS (1667–1754), Anglo-French mathematician. Best known for his theorem in trigonometry; lecturer and teacher of mathematics and natural sciences.

Mørch, Ernst Trier (b. 1908), Danish (naturalized American) physician, anaesthesiologist, educator. Professor of Surgery, and director of anaesthesia, Cook County Hospital, Illinois.

Morgan, Thomas Hunt (1866–1945), American biologist, embryologist, geneticist; early supporter of eugenics, later disillusioned. Late Professor of Biology, California Institute of Technology, Pasadena; author of gene theory; Nobel prize (1933).

Muggeridge, Malcolm (b. 1903), British journalist, broadcaster, author. Lately Editor, *Punch*.

Muller, Hermann J. (1890–1967), American geneticist, student of T. H. Morgan (q.v.). Early interest in eugenics, disillusioned, then revived; worked in USA, Berlin, USSR, Edinburgh; Professor of Zoology, University of Indiana, Bloomington; discoverer of influence of X-rays in producing mutations (1927); Nobel prize (1946).

Nada, John. Pseudonym of John Langdon-Davies (1897–1971), British author, science journalist.

Needham, Joseph FRS (b. 1900), British biochemist, embryologist, physiologist, Sinologist. Master of Gonville and Caius College, Cambridge, since 1966.

Newton, Sir Isaac (1642–1727), English mathematician, physicist, astronomer, philosopher, author. Master of the Mint, London; worked on light and gravitation.

Nietzsche, Friedrich Wilhelm (1844–1900), German philosopher, philologist, poet. Propounded idea of Superman as next stage in evolution of man.

Noyes, John Humphrey (1811–86), American social reformer, preacher. Claimed miraculous healing powers; convinced he was in state of complete sinlessness or perfection; encouraged free love within community; arrested on charge of adultery, but broke bail; led movement for thirty years, then emigrated to Canada when powers failed.

Osborn, Frederick (b. 1889), American businessman, eugenicist, author. Chairman of Executive Committee, Population Council (1930–68); Secretary/ Treasurer, American Eugenics Society (1959–70).

Parker, Dorothy, *née* Rothschild (1893–1967), American author, fashion journalist, wit.

Parkes, Sir Alan S. FRS (b. 1900), British biologist. Chairman, The Galton Foundation, London; Lately Mary Marshall Professor of the Physiology of Reproduction, University of Cambridge.

Pasternak, Boris Leonidovich (1890–1960), Russian writer.

Pauling, Linus Carl FRS (b. 1901), American biochemist, educator. Professor of Chemistry, Stanford University, California; Nobel prize (1954); Peace prize (1962).

Pearson, Karl FRS (1857–1936), British lawyer, statistician, author. Late Galton Professor of Eugenics, University College, University of London.

Peel, John (b. 1930), British sociologist. Reader in Sociology, University of York, England.

Plough, Harold Henry (b. 1892), American biologist. Emeritus Professor of Biology, Amherst College, Massachusetts.

Postgate, John R. (b. 1922), British chemical microbiologist. Professor of Microbiology and Assistant Director, ARC Unit of Nitrogen Fixation, University of Sussex, Brighton.

Potts, D. Malcolm (b. 1935), British obstetrician, author. Medical Director, International Planned Parenthood Federation; Vice-President, Eugenics Society.

Price, Bronson (b. 1905), American physiologist, statistician, human geneticist. Statistician with US Office of Education, Washington DC from 1957.

Price, William (b. 1934), Labour politician, journalist.

Quetelet, Lambert Adolphe Jacques (1796–1874), statistician and astronomer of Brussels, Belgium.

Race, Robert Russell FRS (b. 1907), British geneticist. Lately Director, Medical

Research Council Blood Group Unit, Lister Institute of Preventive Medicine, London.

Reith, John Charles Walsham, Baron (1889–1971), British statesman, engineer, author. First General Manager and Managing Director of British Broadcasting Company; Director General of British Broadcasting Corporation (1926–38).

Renwick, James Harrison (b. 1926), British doctor, geneticist. Reader in Human Genetics, Public Health and Medical Research Council Social Medicine Unit, London School of Medicine and Tropical Hygiene.

Rhodes, Philip (b. 1922), British surgeon. Dean and Professor of Obstetrics and Gynaecology, St Thomas's Hospital Medical School, London.

Rich, Clayton, Dean of Medical School, Stanford University, California.

Richardson, Ken, British biologist, educator, Neurobiological Research Group, The Open University, Milton Keynes, Bucks.

Roberts, Charles (1836–1901), British doctor. Assistant surgeon to Crimean Railway Company during Crimean war; then held local public health appointments in Uxbridge, Middlesex.

Rorvik, David M., American freelance medical writer and journalist.

Rostand, Jean (b. 1894), French author of 50 books on contemporary biological science.

Russell, Bertrand A. W., Earl, OM FRS (1872–1970), British philosopher, author.

Sachs, Bernard (1858–1944), American neurologist. Professor of Clinical Neurology, University of Columbia, New York; discoverer, with Warren Tay (q.v.) of Tay-Sachs disease.

St John-Stevas, Norman A. F. (b. 1929), British Conservative politician.

Sanger, Margaret, née Higgins (1883–1966), American nurse; leader of birth-control movement. Indicted (1915) for sending pleas for birth control through post; organized 1st American Birth Control Conference (1921); founded American Birth Control League.

Sanger, Ruth A. FRS (Mrs R. R. Race) (b. 1918), British (Australian-born) geneticist. Director, Medical Research Council Blood Group Unit, The Lister Institute of Preventive Medicine, London.

Schiller, Ferdinand Canning Scott (1864–1937), British philosopher, author. Late Professor of Philology, Los Angeles University, California.

Schultz, Bruno K. (b. 1901), German anthropologist. Professor and Director of Biological Institute of the Third Reich's Academy for Physical Training.

Schultz, Jack (1904–71), American geneticist, environmental biologist. Institute for Cancer Research, Philadelphia.

Scott, John Paul (b. 1909), American scientist, author. Research Professor of Psychology, Bowling Green State University, Ohio.

Shaw, George Bernard (1856–1950), Irish playwright.

Shettles, Landrum B. (b. 1909), American obstetrician and gynaecologist. Assistant Professor of Clinical Obstetrics and Gynecology, College of Physicians and Surgeons, New York.

Shockley, William B. (b. 1910), English-American physicist. Alexander M. Poniatoff Professor of Engineering Science, Stanford University, California; Nobel prize (1956).

Slater, Eliot T. O. (b. 1904), British psychiatrist. Lately Director of Medical Research Council Psychiatric Genetics Unit, Institute of Psychiatry, London; Vice-President, Eugenics Society.

Slome, John J., British medical practitioner. Author of *Abortion and the Unmarried Mother*.

Spears, David, British biologist, psychologist. Neurobiological Research Group, The Open University, Milton Keynes, Bucks.

Spencer, Herbert (1820–1903), English philosopher, civil engineer, writer. Advocate of Social Darwinism, with extensive influence in USA; spent 1860–1896 elaborating his *Synthetic Philosophy*; declined all academic distinctions.

Steinberg, Arthur G. (b. 1912), American human geneticist. Professor of Biology, Case Western Reserve University, Cleveland, Ohio; also Associate Professor of Human Genetics, Department of Preventive Medicine.

Stephens, Wilson T. (b. 1912), British journalist. Editor, *Field*, London.

Stern, L. Wilhelm (1871–1938), German psychologist, educator, philosopher. Professor of Philosophy, University of Hamburg, Germany (1916–33); Professor, Duke University, North Carolina (1934); pioneer of intelligence tests.

Stevens, S. Smith (b. 1906), American psychophysicist. Professor of Psychophysics, Harvard University, Cambridge, Mass.; previously Director of Psychoacoustic Laboratory, Harvard.

Stevenson, Alan Carruth (b. 1909), British geneticist. Director, Population Genetics Unit, Medical Research Council, Oxford.

Stokes, William Earl Dodge (1852–1926), American capitalist, real estate operator, horse breeder, eugenicist, author.

Stopes, Marie Carmichael (1880–1958), English pioneer of birth control, suffragette, lecturer in palaeobotany. Jointly with Humphrey Verdon Roe, her second husband, founded Mother's Clinic for Constructive Birth Control (first in world) in Holloway, London (1921).

Strigel, Bernhard (*c.* 1460–1528), German painter, principally of portraits; also altarpieces. Worked near Ulm, Germany; style transitional between late Gothic and Renaissance.

Tanner, James M. (b. 1920), British human biologist. Professor and Head of

Department of Growth and Development, Institute of Child Health, University of London.

Tay, Warren (1843–1927), English doctor. Consultant surgeon, Queen's Hospital for Children, also Ophthalmic Hospital and Hospital for Diseases of the Skin, London; discovered, with Bernard Sachs (q.v.), Tay-Sachs disease.

Thoday, John M. FRS (b. 1916), British population geneticist. Arthur Balfour Professor of Genetics, University of Cambridge.

Tietze, Christopher (b. 1908), American demographist and statistician for family planning. Associate Director, Bio-Medical Division, Population Council, New York.

Tizard, Jack P. M. (b. 1919), New Zealand psychologist. Research Professor of Child Development, Institute of Education, University of London.

Treitschke, Heinrich von (1834–96), German historian, with influential nationalistic views.

Tschermak, Erich von Seysenegg (1871–1962), Austrian botanist, plant breeder. Late Professor of High School of Agriculture, Vienna; rediscovered Mendel's law of inheritance (1900).

Twain, Mark. Pen-name of Samuel Langhorne Clemens (1835–1910), American author.

Vallentin, Antonina (b. 1893), Polish biographer, political writer, mostly in German. Emigrated to Paris.

Vasari, Giorgio (1511–74), Italian artist, sculptor, architect, biographer.

Vries, Hugo de FRS (1848–1935), Dutch botanist, plant physiologist and hybridist; 'Father of the Mutation Theory'. Rediscovered Mendel's law of inheritance (1900).

Waddington, Conrad Hal FRS (b. 1905), British geneticist, developmental biologist. Buchanan Professor of Animal Genetics, University of Edinburgh.

Wallace, Alfred Russel (1823–1913), British naturalist, traveller. Published his theory of evolution in 1858.

Watson, James D. (b. 1928), American biochemist. Professor of Biology, Harvard University, Cambridge, Mass.; Nobel prize (1962).

Weismann, August (1834–1914), German biologist, embryologist, geneticist. Professor of Zoology, Freiburg University, Germany; contended that only variations of germplasm, not acquired characters, are inherited.

Wells, Herbert George (1866–1946), English author.

Westoff, Charles F. (b. 1927), American educator, sociologist. Executive director, Commission on Population Growth and the American Future, Washington (1971–2).

Whitney, Leon F. (b. 1894), American biologist, veterinarian, author. Executive Secretary of American Eugenics Society (1924–34). Lately Clinical instructor in pathology, Yale School of Medicine, Connecticut.

Wiggam, Albert E. (1871–1957), American lecturer, journalist, writer, eugenicist. Editorial director, National Newspaper Service, National Science Writers Association, USA.

Wilde, Oscar (1856–1900), Irish playwright, writer.

Wynn, Arthur H. A. (b. 1910), British scientist. Adviser on Standards, Department of Trade and Industry, London (1965–71). Married to Margaret Wynn, British author and student of family policy.

Zola, Emile (1840–1902), French novelist.

GLOSSARY 2

Terms

ACATALASAEMIA – a congenital deficiency of the enzyme catalase in the red blood cells, resulting in *acatalasia,* a rare disease, occurring mostly in Japanese, consisting of gum and mouth infections.

ACCESSORY SEX ORGANS – those which differentiate the sexes but are not crucial to reproduction, e.g. female breasts.

ACHONDROPLASIA – an inherited defect in the cartilage formation of the long bones of the arms and legs producing a form of dwarfism.

ACIDOSIS – an abnormality in the body's acid level, giving rise either to an accumulation of acids (as in diabetes mellitus, q.v.) or an excessive loss of bicarbonate (affecting the kidneys).

ADDISON'S DISEASE – an insufficiency of hormone production by the cortex of the adrenal glands. (Thomas Addison, English physician, 1793–1860, first described it in 1855.)

ADENINE – 6-aminopurine. It is one of the four bases of the DNA molecule.

ADHESION – a holding together by new tissues of normally separate and movable structures, caused by inflammation or injury.

ADRENAL INSUFFICIENCY – a condition caused by the inadequate function of the adrenal gland.

ADRENOGENITAL SYNDROME – a condition, congenital or acquired through excessive growth of tissue, resulting in the abnormal enlargement of some sex organs.

AINU – a Japanese people with an abundance of body hair.

ALBINISM – an inherited (autosomal recessive) disorder characterized by the absence of pigmentation in skin, hair and eyes.

ALOPECIA CONGENITA – baldness, due to absence of hair bulbs at birth.

AMAUROTIC FAMILY IDIOCY – a group of inherited (recessive) familial disorders, characterized by defective intelligence, usually associated with blindness, dementia and early death. (The most common is Tay-Sachs disease, q.v.)

AMINO-ACID – one of a class of organic acids marked by the presence of both an amino NH_2) and a carboxyl (COOH) radical. There are over twenty amino-acids, all of which are essential to metabolism. Only eight of them have to be present in food; the rest can be manufactured by the body.

AMNION – a thin, fluid-filled membranous sac enveloping the foetus. It is filled with *amniotic fluid,* a transparent, almost colourless liquid which serves to protect the foetus, and some of which can be removed by puncturing the sac, a process known as *amniocentesis.* The foetus can be detected by an *amnioscope.*

AMPHETAMINE – a synthetic drug used to relieve nasal congestion, and to stimulate the central nervous system.

ANAEMIA – a deficiency of red blood corpuscles or their haemoglobin content. *Haemolytic anaemia* involves the breakdown of red blood cells and liberation of blood pigment into the bloodstream. *Herrick's* or *sickle-cell anaemia*, an inherited (incomplete dominant) condition characterized by the presence of abnormal sickle- or crescent-shaped red corpuscles, is almost exclusively confined to Negroes. (First described in 1905 by James Herrick, American physician, 1861–1954.) In *megaloblastic anaemia* abnormally large red corpuscles are present in the blood; its most usual form is *pernicious anaemia*, generally occurring after the age of forty, associated with degeneration of the stomach lining. *Splenic anaemia* is characterized by the enlargement of the spleen, gastric haemorrhage and usually cirrhosis of the liver. *Thalassaemia* (or *Cooley's* or *Mediterranean anaemia*) is a hereditary (incomplete dominant) anaemia, with familial or racial incidence, of which the more severe form is characterized by a Mongoloid face, enlargement of the spleen and heart. (T. B. Cooley, American physician, 1871–1945.)

ANENCEPHALY – the absence of brain and the spinal cord.

ANKYLOSING (or RHEUMATOID) SPONDYLITIS – a chronic progressive arthritis suffered by young men, mainly affecting the spine and sacroiliac joints, resulting in fusion and deformity.

ANONYCHIA – the absence of nails.

ANOXIA – a deficiency of oxygen.

ANTIBIOTIC – a substance, either natural or synthetic, which inhibits the growth of or kills micro-organisms.

ANTIBODY – a protein developed by the body which destroys or neutralizes antigens. *Antigens* can be foreign to a body (such as bacteria) or produced within it (such as bacterial toxins). They induce the formation of antibodies.

ARGINOSUCCINICACIDURIA – a hereditary (autosomal recessive) metabolic disorder resulting in the excretion of arginosuccinic acid.

ARTIFICIAL INSEMINATION – the introduction of semen into the uterus by artificial means, not by coitus. *AID*, or *artificial insemination by donor*, involves the use of semen supplied by someone other than the woman's husband. In *AIH*, or *artificial insemination by husband*, a man's semen is introduced artificially into his wife. See also egg transfer.

ASPIRATION, ENDOMETRIAL – see menstrual extraction.

ATAXIA – an inability to co-ordinate movement or muscular activity.

ATRESIA – the absence of a natural opening, or its closure by a membrane.

AUSTRALOPITHECINE – all lower Pleistocene hominids that are not *Homo* (Coon).

AUTISM – a state, becoming apparent about the age of eighteen months, of extreme self-isolation, accompanied by a failure to communicate.

AUTOSOME – a chromosome other than a sex chromosome.

BACK-CROSS – a mating, or the result of a mating, between a hybrid and one of its parent race.

BARBITURATE – a sedative, hypnotic, anaesthetic drug derived from barbituric acid.

BASE – any substance which, when dissolved in water, will provide hydroxyl ions (OH) from its own molecules.

BETA-GALACTOSIDASE DEFICIENCY – a hereditary (recessive) enzyme inadequacy causing (generalized) gangliosidosis (q.v.).

BLASTULA – an early stage in the development of an embryo which consists of a hollow sphere of cells.

BLEB – a small blister on the skin.

BLOOD GROUP – immunologically distinct, genetically determined types of blood.

BRACHYCEPHALY – short or round-headed, the maximum skull breadth being 80% or more of the maximum length.

BRACHYDACTYLY – abnormal shortness of digits.

BRACHYMESOPHALANGY – abnormal shortness of the middle joint of digits.

BROMELIAD – any plant of the Bromeliaceae, a tropical American family.

CAESARIAN (operation or section) – the delivery of a child by cutting through the wall of the abdomen.

CANIDAE – the family to which dogs, foxes, jackals and wolves belong.

CANNULA – a surgical tube for insertion into the body, usually fitting round a pointed instrument (trocar) which is then withdrawn to permit fluid to run out through the cannula.

CAPACITATION – the process occurring after ejaculation which enables the sperm to penetrate and fertilize an ovum.

CARCINOGEN – any substance initiating or encouraging cancer.

CARRIER – one who transmits a hereditary disorder without suffering from it.

CATARACT – a disease of the eye in which the lens becomes opaque, more or less completely obscuring vision.

CEPHALIC INDEX – the ratio of the maximum breadth to the maximum length of the skull, expressed as a percentage.

CEREBRAL PALSY – a malfunction of the brain, usually caused before or during birth, resulting in a non-progressive lack of muscular co-ordination, muscular spasms or paralysis, and speech difficulties. A sufferer is frequently called a spastic.

CERVIX – the neck of an organ, especially the uterus.

CHEDIAK-HIGASHI SYNDROME – an inherited (autosomal recessive) disorder, causing a discolouration of the white corpuscles, enlargement of the spleen and liver, disorders of the lymph nodes and recurrent skin infections. (Moises Chediak, Cuban physician; Otataka Higashi, Japanese physician.)

CHOLERA – a highly infectious, frequently fatal disease, characterized by bilious vomiting and diarrhoea.

CHONDRODYSTROPHIC DWARFISM – an inherited (dominant) defect in the cartilage formation of long bones, causing stunted growth.

CHRISTMAS DISEASE – see under haemophilia.

CHROMOSOME – a small body composed of protein and nucleic acid, of which a constant number for each species is present within the cell nucleus, and which plays a crucial role in the transmission of genetic material to each daughter cell.

CLEFT LIP (or HARE-LIP) – a congenital defect characterized by a fissured upper lip, often combined with cleft palate; more common in females.

CLEFT PALATE – a congenital defect leaving a longitudinal fissure in the roof of the mouth, often combined with cleft lip.

CLONE – a whole stock of individuals derived asexually from a single animal or plant.

CLUB-FOOT – see talipes.

CONGENITAL – present at birth. A *congenital malformation* is a deformity present at birth, either actually (e.g. anencephaly) or potentially (e.g. Huntington's chorea), and usually inherited. The commonest form of *congenital dislocation* is that of the hip where the displacement of one or both joints is potentially crippling. In *congenital flatfoot* there is an abnormal flatness of sole and arch of the foot which may not interfere with its normal function. *Congenital heart disease* is generally caused by the heart's failure to develop properly during embryonic life.

CONSANGUINITY – blood relationship.

CONTERGAN – a brand name for thalidomide.

CONVULSION – a violent, uncontrollable, and usually extensive spasm of muscular contraction.

CRANIAL CAPACITY – the cubic content of the skull.

CRO-MAGNON – a member of the Palaeolithic race of tall, long-skulled but short-faced men associated with the Aurignacian culture (the first skulls of this kind were found at Cro-magnon, Dordogne, France).

CROSS – interbreeding, or the product of such breeding. A hybrid.

CRYSTALLOGRAPHY – the study of the structure, forms and properties of crystalline substances.

CYANATE – a salt or ester of cyanic acid.

CYCLINE – an abbreviation for the tetracycline group of broad-spectrum antibiotics.

CYPROTERONE ACETATE – a steroid which neutralizes testosterone without feminization.

CYSTIC FIBROSIS – an inherited (recessive) disease occurring in children which results in pancreatic insufficiency, chronic pulmonary disease and, in some cases, cirrhosis of the liver.

CYSTINOSIS – a congenital, hereditary (autosomal recessive) disorder resulting in an abnormal concentration of the amino-acid cystine, especially in the kidneys, causing renal insufficiency.

CYTOSINE – oxyaminopyrimidine, a base that is one of the four present in the DNA molecule.

DEAF MUTISM – the inherited (generally autosomal recessive) condition of being both deaf and dumb. The deafness is normally the cause of the mutism but timely treatment can often prevent it being so.

DEOXYRIBONUCLEIC ACID (DNA) – the complex substance of high molecular weight which carries genetic information in coded form.

DES – see diethylstilboestrol.

DEXTROCARDIA – the inherited (possibly autosomal recessive) condition in which the heart is situated on the right side of the chest, instead of the normal left.

DIABETES INSIPIDUS – a disease, either hereditary or caused by a lesion affecting the pituitary gland, characterized by excessive thirst and discharge of urine. More common in the young and very rare compared with d. mellitus. *Nephrogenic diabetes insipidus* (an X-linked recessive trait) results from the congenital failure of the renal tubules to reabsorb water.

DIABETES MELLITUS – a chronic, incurable disorder (often hereditary) characterized by excessive amounts of sugar in the urine.

DIATHERMY – the therapeutic use of high frequency electric current for, in surgical cases, electrocoagulation and cauterization.

DIAZEPAM – a tranquillizer. *Diazepam-hypotonia* is a decrease of vascular tension due to the administration of diazepam.

DIETHYLSTILBOESTROL (DES) – a synthetic oestrogen used as a substitute for natural oestrogenic hormones in the treatment of various female disorders.

DIPHTHERIA – an acute, infectious disease, characterized by a grey membrane spread over the back of the mouth. Commonest in children between 1 and 10.

DISTAVAL – a trade name for thalidomide.

DNA – the common abbreviation for deoxyribonucleic acid.

DOG-FACE – see hypertrichosis universalis.

DOLICHOCEPHALY – long-headedness; the maximum breadth of the skull being less than 75% (or 80%) of the maximum length.

DOUCHE – a current of water directed against a part; a vaginal douche following intercourse is a method of contraception.

DOWN'S SYNDROME (or MONGOLISM or TRISOMY-21) – a type of congenital mental retardation accompanied by physical abnormalities and stunted growth. Called mongolism because of a superficial resemblance of sufferers to Mongoloids. Trisomy means there are three of a chromosome pair (No. 21) per cell instead of two. (First described in 1866 by the English physician, J. Langdon Down, 1826–96.)

DROSOPHILA – a genus of flies which includes the fruit flies that breed in fermenting fruit juices. The species *Drosophila melanogaster* has been extensively used in genetic experiments.

DYSGENIC – unfavourable to race improvement. Contrary to eugenic.

DYSTROPHIA MYOTONICA – a hereditary (autosomal dominant) disease, characterized by lack of normal relaxation of muscles after contraction.

DYSTROPHY – defective nutrition; abnormal development; degeneration. *Apical dystrophy* describes the absence of segments on the fingers. *Muscular dystrophy* is a degenerative muscular disease of which the most common form is *Duchenne's*, or *pseudohypertrophic infantile muscular dystrophy*, a progressive hereditary, sex-linked (male) disease, in which the muscle cells are scanty, the remainder having been replaced by fat, giving the appearance of bulging muscles, and resulting in a lack of power. (G. B. A. Duchenne, French neurologist, 1806–75.)

ECTRODACTYLY – the congenital absence of one or more digits.

EGG TRANSFER – in *ETD*, or *egg transfer from donor*, the ovum is donated by another woman and then fertilized with sperm from her husband. In *ETDD*, or *egg transfer from donor plus sperm transfer from donor*, the ovum donated by another woman is fertilized by semen from a man other than the would-be father. In *ETW*, or *egg transfer from the wife*, the woman herself provides the ovum for insertion. See also artificial insemination.

EMBOLISM – the presence of an obstruction in a blood vessel, either a blood clot or, as in *air embolism*, air in the bloodstream.

EMBRYO – the early or developing stage of any organism. For humans the embryonic period is from about the second week after conception (some say earlier) to about the end of the second month. Thereafter the embryo becomes a foetus.

ENZYME – an organic compound, often a protein, produced by living cells which acts as a catalyst in causing or accelerating specific chemical reactions, e.g. the conversion of starch into glucose by the salivary enzymes ptyalin and maltase.

EOCENE – the second oldest Tertiary period. (It had been the oldest but in 1874 the Palaeocene period was inserted between it and the Cretaceous.)

EPICANTHUS – a fold of skin above the inner angle of the eye, a Mongoloid characteristic.

EPIDERMIS – the external layer of an animal's skin.

EPIDIDYMIS – a convoluted tube attached to each testis in which spermatozoa are stored.

EPILEPSY – a nervous disorder affecting males more than females, generally characterized by convulsions, and sometimes causing the sufferer to fall down and lose consciousness.

EPILOIA (or TUBEROSE SCLEROSIS) – a rare hereditary (dominant) disease, due to a developmental abnormality of the brain, in which the main features are mental deficiency, epilepsy and sebaceous tumour-like growths on the skin, especially of the face.

EUGENICS – the branch of biological science studying the inherited characteristics of humans with particular interest in their improvement. *Negative eugenics* is the prevention and *positive eugenics* the encouragement of the breeding of certain types.

EUTHANASIA – the practice of painless killing, especially in cases of incurable suffering.

FACIAL INDEX – the ratio of the breadth to the length of the face, expressed as a percentage.

FALLOPIAN TUBE (or uterine tube or oviduct) – the duct through which ova pass from the ovary to the uterus. (Gabriel Fallopius, Italian anatomist, 1523–62.)

FOETAL BRADYCARDIA – abnormal slowness of heartbeat of the foetus.

FOETAL PARESIS – slight paralysis of the foetus.

FOETUS – the human organism from about the third month of development to birth.

FOLLICLE – any small sac-like structure or cavity.

GALACTOSAEMIA – an inherited (autosomal recessive) error of metabolism noticeable shortly after birth, where the absence of the enzyme necessary for converting galactose to glucose results in a failure to thrive in infancy, vomiting, diarrhoea, jaundice and mental retardation.

GALL-STONE – a deposit of cholesterol, and sometimes bile salts, in the gall bladder or bile duct.

GAMETE – a sexual reproductive cell; an egg cell or sperm cell.

GANGLIOSIDOSIS – any degenerative disease of the central nervous system involving an abnormality of gangliosides, a type of fatty acid found in nerve cell membranes (as in Tay-Sachs disease, q.v.).

GASTROENTERITIS – the inflammation of the stomach lining and intestines, causing vomiting, diarrhoea and cramp.

GAUCHER'S DISEASE – an inherited (autosomal recessive) disorder resulting in chronic enlargement of the spleen and liver. (Philippe Gaucher, French physician, 1854–1914.)

GENE – the basic unit of heredity, capable of self-reproduction, which usually occupies a definite place on a chromosome.

GENE-FLOW – the transfer of one or more genetic characters from one population of an organism to another population of the same organism.

GENE POOL – the total content of genetic characteristics within a population.

GENETICS – the biological science that is concerned with studies of natural differences and resemblances among related organisms.

GENITAL – pertaining to the reproductive organs.

GERMPLASM – the reproductive tissue as distinct from the non-reproductive tissue.

GLAND – a single structure of cells taking substances from the blood and secreting them either down a duct or into the bloodstream, in a form which the body can use or eliminate. The *pituitary gland* is a ductless gland attached to the base of the brain; its posterior lobe secretes hormones affecting the kidneys, muscular control and reproduction; its anterior lobe secretes hormones regulating most other ductless glands. The *prostate gland* is an organ at the neck of the male bladder which secretes a fluid that forms part of a male's semen. *Sebaceous glands* excrete an oily substance on to the skin, generally via the hair follicles.

GLAUCOMA – an eye disorder caused by excessive fluid pressure within the eyeball, resulting in gradual decrease of vision.

GLOBULIN – one of a group of simple proteins that coagulate under heat and are insoluble in water.

GLUCOSE-6-PHOSPHATE DEHYDROGENASE (or G6PD) DEFICIENCY – a congenital inherited, sex-linked (male) red-cell defect, in which very low levels of this enzyme are associated with a tendency towards haemolytic anaemia if certain drugs (e.g. sulphonamides) are administered.

GLUTAMIC ACID – an amino-acid and the only one metabolized by the brain. Used in the treatment of epilepsy.

GLYCOGEN STORAGE DISEASE – one of a group of inherited (autosomal recessive) disorders caused by a deficiency of one or more of the enzymes essential for glycogen (animal starch) synthesis or degradation.

GOITRE – an enlargement of the thyroid gland, characterized by a swelling at the front of the throat, sometimes resulting from iodine deficiency in the diet.

GONADOTROPHIC HORMONE (or GONADOTROPHIN) – a hormone produced by the anterior lobe of the pituitary gland which stimulates the testes and ovaries.

G6PD DEFICIENCY – see glucose-6-phosphate dehydrogenase deficiency.

GUANINE – 2-Aminohypoxanthine, a purine which is one of the four bases of the DNA molecule.

HAEMOPHILIA – a hereditary (X-linked recessive) tendency to excessive bleeding, even after slight injury, due to lack of a clotting factor; only affects males. *Haemophilia B*, or Christmas disease (the first reported case was S. Christmas, an English boy) is distinguishable from *Haemophilia A* (classical haemophilia) by blood tests; a different lack is involved.

HALF-BROTHER, -SISTER – related by one parent only but with genes in common. For step-relationships there are no genes in common.

HARE-LIP – see cleft lip.

HELIX – a coil or spiral; the structure of DNA (q.v.) is a *double helix* with two helices intertwined.

HERNIA – the protrusion of an organ or tissues through the wall of the cavity which contains them. With *femoral hernia* the intestines descend through the femoral ring. With *inguinal hernia* (four-fifths of the total) the sac containing the intestine protrudes at the inguinal opening.

HETEROCHROMIA IRIDIS – the heritable condition of having each iris a different colour.

HETEROSIS – cross-fertilization, or the improvement of a species by outbreeding; hybrid vigour.

HETEROZYGOTE – a zygote (or fertilized ovum), or the animal or plant developing from it, formed by the union of two gametes dissimilar for a particular gene.

HEXADACTYLY – the possession of six digits instead of five.

HEXOSAMINIDASE – a crucial enzyme in sugar metabolism whose absence or inadequacy gives rise to Tay-Sachs disease (q.v.).

HOMOCYSTINURIA – an inherited (autosomal recessive) disorder characterized by incompletely dislocated lenses (developing after the age of ten), thromboembolisms and mental retardation.

HOMOGAMY – inbreeding; the breeding of like with like.

HOMOZYGOTE – a zygote (or fertilized ovum), or the animal or plant developing from it, formed by the union of two gametes similar for a particular gene.

HORMONE – a substance secreted by an endocrine gland which has a specific effect on the activity of another organ.

HUNTER'S SYNDROME – mucopolysaccharidosis II: a hereditary (X-linked recessive) disorder with similar characteristics, though in a milder form, to Hurler's syndrome (q.v.) which is the commoner. (Charles Hunter, Canadian physician, 1872–1955.)

HUNTINGTON'S CHOREA – a rare inherited (dominant) motor disorder characterized by jerky, spasmodic movements, leading to mental deterioration and death within a few years; congenital but not apparent until middle age. (George Huntington, American physician, 1850–1916.)

HURLER'S SYNDROME (or LIPOCHONDRODYSTROPHY) – mucopolysaccharidosis I: a congenital inherited (autosomal recessive) abnormality in the skeletal bones and cartilage, leading to deformity, possible mental deficiency and defective vision. (Gertrud Hurler, Austrian paediatrician, described it in 1919.)

HYBRID – the offspring of parents differing in species or variety; a heterozygote.

HYDROCEPHALUS – an excessive amount of fluid within the skull, resulting in abnormal enlargement of the head, atrophy of the brain and limited mentality.

HYDROLYSATE – a compound produced by chemical decomposition of a substance in water; *protein hydrolysate* is a mixture of amino-acids prepared by breaking down proteins with acid, alkali or enzyme, for use in special diets.

HYPER – too much, as against *hypo* which is too little.

HYPERAMMONAEMIA – an inherited (autosomal recessive) disorder causing an abnormally high level of ammonia in the blood resulting, particularly after a protein-rich meal, in vomiting, lethargy, indistinct speech and mental inadequacy.

HYPERBILIRUBINAEMIA – a hereditary (autosomal recessive) disorder resulting in an excessive amount of bilirubin (bile's yellow pigment) in the blood; a severe and prolonged jaundice sometimes affecting a premature baby, and resulting in mental retardation.

HYPERCALCAEMIA – an excessive amount of calcium in the blood, upsetting the calcium/phosphate ratio, and resulting in poor bones.

HYPERDACTYLY – the possession of more than five digits on a limb.

HYPERTRICHOSIS UNIVERSALIS (or DOG-FACE) – a congenital disorder characterized by an excessive growth of hair over more of the body than normal.

HYPERVALINAEMIA – a hereditary metabolic disorder resulting in abnormal quantities of the amino-acid valine (essential for infant growth and adult nitrogen balance) in blood and urine, characterized by failure to thrive in infancy and mental retardation.

HYPOGLYCAEMIA – a deficiency of glucose in the blood, sometimes caused by an overdose of insulin, characterized by hunger, sweating and, on occasion, convulsions.

HYPOPHALANGY – the absence of a segment of a digit.

HYPOTHERMIA – subnormal body temperature.

ICTHYOSIS – a skin disease, generally hereditary (dominant) and persisting throughout life, in which the skin surface is very rough and cracked, with the appearance of fish scales.

IDIOT – strictly an individual with the lowest grade of intellect, e.g. an adult with the mental age of 2 or an IQ not exceeding 25. See imbecile.

ILEUS – paralysis of a segment of the intestines resulting in an obstruction.

IMBECILE – strictly an individual with a mental age between 2 and 7, or an IQ between 25 and 50. See idiot.

IMMUNOLOGY – the science dealing with specific mechanisms, such as antibody-antigen reactions, used by an organism against foreign material. These mechanisms can result in increased resistance (as in immunity) or in damaging reactivity (as in allergy).

IMPLANTATION – the embedding of a developing embryo in the uterus.

INSULIN – the hormone secreted from the islet of Langerhans in the pancreas that maintains blood sugar levels.

INTELLIGENCE QUOTIENT (IQ) – the ratio of mental to chronological age expressed as a percentage.

INTRAUTERINE DEVICE (IUD) – any mechanical contrivance inserted into the womb to prevent implantation or growth of an embryo.

IONIZING RADIATION – radiation which knocks electrons from atoms, thereby leaving ions behind.

JAUNDICE – a disease in which an excess of bile pigment causes yellowing of the eyes, skin, etc., accompanied by digestive disturbance (with dark urine, pale faeces) and slow pulse.

KERNICTERUS – the infiltration of bile into the basal nuclei of the brain, with toxic degeneration of the nerve cells, occurring in jaundiced babies in the first week of life and frequently fatal.

KIDNEY-STONE – an abnormal concretion (solid coalescence), usually of mineral salts, in a kidney.

LACTOSE – the sugar present in milk.

LAPAROSCOPY – examination of the interior of the abdomen by means of an instrument introduced through a small incision in the abdominal wall.

LEPTOSPIROSIS – a disease carried by bacteria of the genus Leptospira, giving rise to a brief fever, but sometimes to greater complications associated with the kidneys or liver.

LESCH-NYHAN SYNDROME – a hereditary (X-linked recessive) disease of male children, characterized by abnormal amounts of uric acid in the blood, resulting in mental retardation and cerebral palsy. (M. Lesch, American paediatrician, b. 1939; W. L. Nyhan, American paediatrician, b. 1926.)

LESION – any change in the structure of a tissue or organ due to injury or disease and usually resulting in the impairment of normal function.

LIGNOCAINE – a local anaesthetic widely used in dentistry.

LIPID – one of a large group of organic compounds, including fats and waxes, found in living cells.

LOBSTER CLAW – a congenital hereditary (dominant) deformity of the hand or foot in which two digits are abnormally large, probably the first and fifth, and the rest are absent.

MALAR – pertaining to the cheek bone area.

MANIC-DEPRESSIVE PSYCHOSIS – a mental disorder (possibly hereditary) alternating between periods of intense depression and excitement.

MAPLE SYRUP URINE DISEASE – a hereditary (usually recessive) disorder caused by an enzyme deficiency, characterized by the smell of the urine shortly after birth and, at its most severe, by vomiting, seizures and death.

MARFAN'S SYNDROME – a hereditary (autosomal dominant) disorder of connective tissue, bones and muscles, characterized by a stooping, unsteady gait, unusually long legs and fingers, and often by deformities of the chest cage and eye lenses. (B-J. A. Marfan, French paediatrician, 1858–1942.)

MARINESCO-SJÖGREN SYNDROME – a congenital inherited (autosomal recessive) cerebellar disorder, accompanied by mental retardation, cataract and minor skeletal anomalies. (Georges Marinesco, Rumanian neurologist, 1864–1938; Tage Sjögren, Swedish paediatrician, 1859–1939.)

MATING – pairing up with the aim of procreation. *Assortative mating* is the choice of a partner with similar characteristics. *Disassortative mating* is the deliberate choice of a mate with dissimilar characteristics. *Random mating* is the selection of a partner without regard to such likenesses.

MEASLES (or RUBEOLA) – an acute infectious viral disease, mostly occurring in children, characterized by catarrh and fever, followed by small red spots with bluish-white centres. *German measles* (or *rubella*) is a different acute infectious viral disease, considerably less severe than measles, characterized by catarrh, slight fever and enlarged neck glands, followed shortly by small spots. German measles can have a teratogenic effect upon the foetus when suffered by a woman within the first four months of pregnancy, particularly if contracted during the first four weeks.

MELANIN – the dark pigment in hair, eyes and skin.

MENARCHE – the onset of menstruation, occurring normally between the tenth and seventeenth years.

MENINGOCELE and MENINGOMYELOCELE – two kinds of protrusion through the skull or spinal column of the membranes (meninges) that envelop the brain and spinal cord. Can be a result of spina bifida (q.v.).

MENSTRUAL EXTRACTION (or ENDOMETRIAL ASPIRATION in USA) – the drawing out by suction of the lining mucosa of the uterus.

MENSTRUATION – the discharge of a bloody fluid from the uterus occurring at fairly regular intervals from puberty to menopause. It results from the failure of a discharged ovum to become fertilized.

MESOLITHIC – between the Palaeolithic and the Neolithic, starting about 8,000 BC and ending either 2,000 years later or much more recently according to area.

MESSENGER-RNA – the ribonucleic acid carrying protein information from the DNA to the ribosomes.

MESTEROLONE – a steroid promoting male secondary sex characteristics.

METABOLISM – the process of making food into complex elements and also of transforming these elements into simpler ones in the production of energy. A product of this process is known as a *metabolite*.

METACARPO-PHALANGEAL JOINT – a joint linking a digit to the hand.

METACHROMATIC LEUKODYSTROPHY – an inherited (autosomal recessive) degenerative disease due to an enzyme deficiency, characterized by colour change in cerebral white matter, resulting in progressive neurological disorder.

METHYLMALONIC ACIDURIA – an inborn (autosomal recessive) error of metabolism resulting in chronic metabolic acidosis.

MICROCEPHALY – abnormal smallness of head and brain.

MICROMELIA – abnormal smallness of the limbs (as suffered by thalidomide victims).

MICROPHTHALMIA – abnormally small eyes.

MISCEGENIST – one who practises interbreeding, especially between different races.

MOLE – a congenital mark or small permanent spot on the skin, often pigmented and hairy.

MOLECULAR BIOLOGY – the branch of science dealing with the relationship between the properties of specific molecules and the characteristics of living things in which these molecules occur.

MONGOLISM – see Down's syndrome.

MONGOLOID – a member of the Mongolian race, or type.

MORVAN'S SYNDROME – a chronic progressive disease of the spinal cord, resulting in painless lesions on the ends of digits, particularly the fingers. (Augustin M. Morvan, French physician, 1819–97.)

MUCOPOLYSACCHARIDOSIS – one of a group of inborn metabolic errors involving mucopolysaccharide, a kind of sugar, which include Hunter's syndrome (q.v.), Hurler's syndrome (q.v.).

MULTIPLE EXOSTOSES – a common hereditary (dominant) disorder of connective tissue in which bony growths project from the surface of a bone, often resulting in the ossification of muscular attachments.

MUTATION – a biological variation, suddenly appearing and transmitted to offspring. A *mutant* is the form arising through mutation, and a *mutagen* is any agent that causes a mutation.

MUTISM – the inability or failure to speak.

MYASTHENIA GRAVIS – a disorder characterized by muscular weakness and fatigue after little exertion, possibly due to a deficiency of acetylcholine which is crucial to muscular contraction.

MYCIN – denotes that a substance, usually an antibiotic, has been derived from a fungus.

MYOPIA – shortness of sight.

NAEVUS – a pigmented area on the skin, usually congenital; a birthmark.

NASAL INDEX – the ratio of the breadth to the length of the nose, expressed as a percentage.

NATURAL SELECTION – a process by which the evolution of a species takes place, with those best adapted to a particular environment surviving best. (The phrase was first used by Charles Darwin.)

NEANDERTHAL – descriptive of a Palaeolithic species of man, whose remains were found in the Neander valley, Germany, in 1856.

NEOLITHIC – pertaining to the last period of the Stone Age which started about 6,000 BC in some areas and more recently in others.

NEONATE – a child in the first few weeks of life.

NEOTENY – the retention in adulthood of immature characteristics.

NEURON – the structural and functional unit of the nervous system; a nerve cell.

NIEMANN-PICK DISEASE – a hereditary (autosomal recessive) disorder of early infancy, usually fatal before the third year, which is due to an enzyme deficiency and characterized by enlargement of the liver and spleen, anaemia and progressive deterioration. (Albert Niemann, German paediatrician, 1880–1921; Ludwig Pick, German physician, 1868–1935.)

NUCLEOTIDE – the basic structural unit of nucleic acid.

NYSTAGMUS – a disorder of the eye that is sometimes congenital and is characterized by involuntary jerky movements.

OESTROGEN – any substance, whether synthetic or natural, which initiates oestrus and the development of secondary sex characteristics. Oestrogenic hormones are formed naturally in the ovaries and the placenta.

OESTRUS – the period when animals are 'on heat' and ready to mate. The oestrous cycle is the sequence of changes in the uterus and ovary.

OOCYTE – an egg cell before it becomes a mature ovum.

ORNITHINE-α-KETOACID TRANSAMINASE DEFICIENCY – the shortage or lack of this particular enzyme.

OROTICACIDURIA – a rare inherited (recessive) disorder of pyrimidine metabolism resulting in a reduced number of white corpuscles.

OSTEOGENESIS IMPERFECTA – a congenital inherited (dominant) bone disease causing the bones to fracture easily, particularly in adolescence. There are various forms and some fractures may even occur before birth.

OVARY – the female egg-producing gland.

OVULATION – the discharge of an ovum.

OVUM – the female gamete; an egg cell.

PAEDIATRICS – the branch of medicine dealing with the growth and development of children.

PALAEOLITHIC – pertaining to the earliest Stone Age which started with the most primitive stone implements and ended in 8,000 BC.

PANDEMIC – descriptive of a disease affecting a whole people/population; epidemic over a very wide area.

PANMIXIS – unrestricted interbreeding; random mating.

PARALYSIS – loss of muscle function or of sensation due to nerve injury.

PARKINSON'S DISEASE – a chronic palsy affecting the elderly, characterized by muscle rigidity and trembling hands. (James Parkinson, English physician, 1755–1824.)

PARTHENOGENESIS – the development of a new organism from an unfertilized ovum.

PEDIGREE – a record of ancestry.

PEDUNCULATED POSTMINIMUS – a small stalk-like growth.

PENICILLIN – one of a large group of antibiotic substances derived from cultures of *Penicillium* fungi.

PEUTZ-JEGHER'S SYNDROME – an inherited disorder characterized by polyps in the small intestine. Their bleeding gives rise to anaemia, accompanied by melanin pigmentation of hands, mouth and feet. (J. L. A. Peutz, Dutch physician; H. Jegher, American physician, b. 1904.)

PHALANX – a bone of a digit, or that part of the digit.

PHENYLALANINE – an enzyme essential for human nutrition, its excess in the blood giving rise to phenylketonuria (q.v.).

PHENYLKETONURIA (PKU) – an inherited (autosomal recessive) metabolic disorder in infants, in which there is an inability to break down phenylalanine; if not immediately treated it results in mental deficiency.

PHTHISIS – tuberculosis, especially of the lungs; any wasting disease.

PHYSIOLOGY – the branch of biological science investigating the processes of life in animals and plants.

PIGMY – dwarf groupings of people living in equatorial Africa. (Greek pygme $= 13\frac{1}{2}$ in, measured from elbow to knuckles.)

PINNA – auricle or external ear.

PLACENTA – the vascular organ uniting an unborn child to its mother's uterus and through which it obtains nutrition.

PLASMA – the liquid part of blood or lymph.

PLASMODIUM FALCIPARUM – the species of protozoa that causes the falciparum form of malaria.

PLEISTOCENE – the final epoch of the Tertiary period which lasted from the Pliocene until the beginning of the Quaternary period a million years ago.

PLIOCENE – the geological period between the Miocene and the Pleistocene.

POLIOMYELITIS – a virus infection attacking and often destroying the motor nerve cells, especially of children. It is characterized by fever, motor paralysis and muscular atrophy; often resulting in permanent deformity and sometimes in death.

POLYANDRY – marriage of one woman with more than one man. To be contrasted with polygyny, the marriage of one man with more than one woman.

POLYDACTYLY – the condition of having extra digits.

POLYGAMY – the state of marriage with more than one person at a time.

POLYGENIC – pertaining to, or caused by, several genes.

POLYMORPHISM – the occurrence of several types or forms of the same species. Male and female are an example.

POLYP – a projecting growth from a membraneous surface.

PORPHYRIA – a congenital inherited error of metabolism resulting in the excretion of purplish-coloured pigment (porphyrin) in the urine, accompanied by great pain, abnormal skin pigmentation and photosensitivity. *Acute intermittent porphyria* (dominant) is characterized by excessive excretion of porphyrin, acute abdominal pain, neurological disturbance and sensitivity to light. *Erythropoietic porphyria* is a rare (recessive) error of porphyria metabolism, appearing in infancy or early childhood, characterized by sensitivity to light, blisters, pigmentation and anaemia. *Variegate porphyria* (also known in South Africa as van Rooyen hands) (dominant) is characterized by skin lesions and acute attacks of jaundice and abdominal colic.

PROGESTERONE – the female sex hormone that prepares the uterus for the fertilized ovum and maintains pregnancy. *Progestogen* is any substance having a similar effect to progesterone.

PROTEIN – one of a group of complex nitrogenous substances of high molecular weight, essential to the growth of tissue in all living cells.

PROTOPLASM – the physicial basis of life; the only known form of matter in which life is manifested.

PROTOZOA – unicellular organisms.

PSYCHIATRY – the study and treatment of mental and nervous disorders.

PSYCHOLOGY – the study of mind and behaviour.

PUBERTY – the beginning of sexual maturity.

QUICKENING – the movement of the foetus in the womb, generally noticeable by the mother around the eighteenth week.

RABIES (or HYDROPHOBIA) – a viral disease transmitted by the bite of an infected animal, resulting in convulsions and delirium, leading to death.

RACE – a division of mankind sharing and transmitting distinct characteristics.

RADIOACTIVE – any unstable substance emitting either atomic particles or gamma radiation.

REFSUM'S DISEASE – an inherited disorder characterized by visual disturbance, inflammation of the nerves and cardiac damage. (Sigvald Refsum, Norwegian neurologist, b. 1907.)

RETINOPATHY – any disorder of the retina.

RETROLENTAL FIBROPLASIA – the development of an opaque fibrous membrane behind the lens, often due to the administration of excessive oxygen to premature babies.

RH-ERYTHROBLASTOSIS – a fairly rare congenital blood disorder of the newborn, caused by the lack of the rhesus factor in the mother and its presence in the baby, leading to destruction of the infant's red blood cells. Blood replacement may be necessary, and is sometimes carried out before birth.

RHESUS FACTOR – a substance present in the blood of 85% of the population.

RIBONUCLEIC ACID (RNA) – a nucleic acid present in all living cells, crucial to cell development and protein synthesis.

RIBOSOME – a distinct particle or granule of a cell on which proteins are assembled and synthesized.

RICKETS – a disease of children, caused by vitamin D deficiency, resulting in defective bone growth.

RNA – see ribonucleic acid, messenger-RNA, transfer-RNA.

RUBELLA – see measles.

RUBEOLA – see measles.

SALICYLATE – any salt of salicylic acid, a white crystalline acid derived from phenol. The salt is used in aspirin and as a preservative and flavouring agent.

SANDHOFF'S DISEASE – a rare form of gangliosidosis (q.v.) with characteristics similar to Tay-Sachs disease (q.v.).

SCHIZOPHRENIA – a mental disorder, characterized by the inability to distinguish between reality and delusion.

SCROTUM – the pouch containing the testicles.

SECONDARY SEX CHARACTERISTIC – one of the differences not concerned with reproduction between males and females, e.g. hair, muscle and fat distribution, and voice.

SEMEN – the fluid produced by the male reproductive organs that carries spermatozoa.

SEMINAL VESICLE – one of the two sac-like structures located behind the bladder which secrete a thick viscous fluid forming part of male semen.

SEROLOGY – the study of serum, the colourless liquid remaining after blood has clotted.

SEX – the fundamental distinction, relating to reproduction within a species, of two forms, male and female, each of which supplies its type of gamete, sperm or ovum, to create the united cell that initiates any offspring.

SEXUAL SELECTION – the choosing for mating of a member of the opposite sex because of certain characteristics. If these are sufficiently popular they will thereafter increase in the population and help to determine the direction of evolution.

SITUS INVERSUS – a congenital disorder, sometimes inherited (autosomal recessive), causing the internal organs to be located on the contrary side of the body.

SMALLPOX – a contagious, febrile disease, characterized by skin pustules; frequently fatal.

SOMA – all the body cells, except the germ cells of reproduction. In Aldous Huxley's *Brave New World* soma is 'the perfect drug ... euphoric, narcotic, pleasantly hallucinant'.

SPASTIC TETRAPLEGIA – the inherited (possibly X-linked) condition of paralysis of all four limbs.

SPERMATOZOA – male germ cells. They are formed within the testes.

SPINA BIFIDA – a congenital defect in which two parts of the bony spinal canal fail to unite perfectly at the embryo stage permitting subsequent protrusion of the cord membranes and giving rise to deformity.

SPINAL CANAL – a tube enclosing the spinal cord formed by the neural arches of the vertebrae.

STENOSIS – the constriction of a tube or passage. *Mitral stenosis* is an obstruction to the flow of blood through the heart's mitral valve. *Pyloric stenosis* is an obstruction (usually congenital) between the stomach and the duodenum. *Infantile pyloric stenosis* is characterized by violent vomiting.

STEP – affinity by another marriage or mating. No genes are shared with a step-relation.

STEROL – one of a large class of complex alcohols, such as cholesterol. A *steroid* is a general term applied to any substance chemically related to sterols, such as sex hormones and bile acids.

STUPOR – a relatively unconscious state, in which there is still a response to stimuli.

SUBCUTANEOUS – pertaining to the loose cellular tissue beneath the skin.

SUPRAORBITAL RIDGE – the curved projecting margin of the frontal bone above the eye.

SYMPHALANGISM – the inherited condition of stiff fingers, or the fusion of contiguous phalanges.

SYMPTOM – a characteristic indication of the existence of a state, especially a disease. In general, symptoms are what a person feels, whereas signs are what a doctor sees.

SYNDACTYLY (or WEBBING) – the possession of two or more digits joined together.

SYNDROME – a characteristic pattern or group of symptoms.

TALIPES – any one of a variety of deformities of the foot, generally congenital' such as club-foot or *equinovarus*, which both involve a rotation of the foot.

TAY-SACHS DISEASE – an inherited (recessive) infantile form of amaurotic idiocy (common among American Jews) caused by a deficiency of the enzyme hexosaminidase. It results in a steady deterioration of sight and of mental and physical abilities from the age of a few months until death by the age of four. (Warren Tay, English physician, 1843–1927; Bernard Sachs, American neurologist, 1859–1944.)

TERATOGENIC – producing abnormal structures in an embryo.

TERMINATION – the ending of a pregnancy by human intervention.

TESTICLE, TESTIS – one of the pair of male reproductive glands producing spermatozoa and testosterone after sexual maturity.

TESTOSTERONE – the principal male sex hormone, a steroid secreted by the testes.

THALASSAEMIA – see under anaemia.

THALIDOMIDE – a sedative drug causing severe malformation of the foetus when taken by the mother in early pregnancy.

THOMSEN'S DISEASE (or MYOTONIA CONGENITA) – an inherited (dominant) muscular disease appearing in early childhood, characterized by muscular spasms and enlargement of skeletal muscles. (Asmus Julius Thomsen, Danish physician, 1815–96.)

THROMBOCYTOPENIA – an inherited (possibly X-linked) condition, where there is a low number of blood platelets caused by a general tendency to bleed spontaneously from small vessels, especially in the skin and mucous membranes.

THROMBOSIS – the formation of a clot in circulating blood.

THYMINE – 5-Methyluracil, a pyrimidine. One of the four bases of the DNA molecule.

TRANSFER-RNA – the ribonucleic acid that carries amino-acids to the ribosomes for processing.

TRISOMY-21 – see Down's syndrome.

TRYPANOSOMIASIS – any of the several diseases caused by trypanosoma, the genus of parasitic protozoa, such as sleeping sickness in Africa and Chagas' disease in South America. Transmitted to man by biting insects.

TUMOUR – an enlargement or swelling; a spontaneous growth of tissue forming an abnormal mass.

TWIN – one of a pair of offspring developed within the uterus at the same time. The pair may be identical and created from a single ovum, or dissimilar and created from two different ova fertilized by two different sperm.

ULCER – an open sore, sometimes forming pus. A *duodenal ulcer* and a *gastric ulcer* are the same condition located in different places, the former on the mucosa of the duodenum, the latter within the stomach; both are classed together as *peptic ulcers*.

UMBILICUS – the navel.

URETHRA – the canal through which urine is discharged from the bladder. In the male it also serves as the passage for semen.

UTERUS – the organ in which the embryo and foetus are nourished and contained; womb.

VACCINATION – the conferring of immunity or increased resistance to a disease by inoculation.

VALINE – an amino-acid constituent of many proteins, essential for normal growth in infants and for nitrogen balance in adults.

VAN ROOYEN HANDS – see under porphyria.

VAS DEFERENS – the duct carrying sperm from the testis. Both vasa are severed in vasectomy.

VASECTOMY – the excision of a segment of the vas deferens, particularly to produce male sterility.

VIRILIZATION – the induction or development of masculine characters in the female.

VIRUS – any one of a large group of infective agents that can only reproduce in living tissue; responsible for diseases such as measles, poliomyelitis, influenza, chicken pox, smallpox, etc.

X-RAY – an electromagnetic wave of very short wavelength which can both penetrate matter opaque to light-rays and act on photographic film.

XERODERMA PIGMENTOSUM – a rare inherited (recessive) skin disease, starting in childhood, characterized by pigment discolouration, ulcers, muscular atrophy and death.

YELLOW FEVER – an acute viral tropical disease, transmitted by the bite of a female mosquito, characterized by fever and jaundice.

ZOONOSIS – an animal disease that can be communicated to man.

ZYGOTE – the single cell formed by the union of a male and a female gamete.

APPENDIX 1

Some drugs for which there is satisfactory evidence of a
possible harmful effect on the foetus (the list being first printed
in the *Medical News-Tribune*)

Taken by mother during pregnancy	Effect on child
Thalidomide, tetracycline, streptomycin	Micromelia
Thalidomide, phenmetrazine, imipramine, tolbutamide, dichlorophenol (defoliant), trimethoprim, anti-mitotics (e.g. methotrexate, cyclophosphamide, busulphan). Ethionamide (TB). Troxidone. Salicylates	Mixed anomalies
Phenytoin and other anticonvulsants	Cleft palate
Phenothiazine, sulphonamides, salicylates, novobiocin, promethazine, chlorpromazine, sparine, vitamin K excess, nitrofurantoin (haemolysis). Corticosteroids, Ristocetin, Chlorothiazide	Jaundice. Hyperbilirubinaemia, risk of kernicterus
Anti-epileptic drugs (phenobarbitone, phenytoin, primidone), quinine, tolbutamide, salicylates, chlorothiazide, chloroquin	Coagulation defects (thrombocytopenia, etc.) bleeding
Anticoagulants (coumarins)	Haemorrhage
Dicophane (DDT)	Tumours
Mercury (Minemata disease)	Cerebral palsy

Taken by mother during pregnancy	*Effect on child*
Lead	Abortion
Streptomycin, quinine, vancomycin, kanamycin, gentamycin, thalidomide	Deafness
Chlorpromazine, chloroquin	Blindness (retinopathy)
Antidiabetic drugs, insulin, tolbutamide	Hypoglycaemia
Penicillin	Sensitization (to pencillin)
Tetracycline	Yellow teeth, enamel defects
Chloramphenicol	Circulatory collapse
Reserpine	Nasal congestion, foetal death, lethargy
Hexamethonium	Ileus
Iodine (e.g. in cough medicine), anti-thyroid drugs, lithium, phenylbutazone	Goitre
Radioactive iodine	Irradiation
Ammonium chloride	Acidosis
Bromides	Rashes
Isoniazid	Lethargy
Corticosteroids	Low birthweight. Cleft palate. Stillbirths. Possible adrenal insufficiency. Foetal distress
Androgens and progestogens, steroids, stilboestrol, contraceptive pill. Gonadotrophins	Virilization. Multiple pregnancy. Carcinoma of vagina in daughters.

Taken by mother during pregnancy	Effect on child
Vitamin A excess	Cleft palate, syndactyly, other anomalies
Vitamin D excess	Hypercalcaemia
Any vitamin excess	Vitamin imbalance, infant to need more than usual dose
Anaesthetics, analgesics, diazepam, pethidine	Respiratory depression. Diazepam-hypotonia, hypothermia
Barbiturates	Respiratory depression
Local anaesthesia for mother (e.g. spinal epidural)	Foetal bradycardia
Magnesium sulphate (for eclampsia)	Lethargy
Muscle relaxants	Foetal paresis
Oxytocin	Risk of cerebral damage

The hereditary diseases that can be diagnosed in the uterus (according to a report presented to the World Health Organization)

Amino-acid disorders
  Arginosuccinicaciduria
  Cystinosis
  Homocystinuria
  Hyperammonaemia
  Hypervalinaemia
  Maple syrup urine disease
  Methylmalonic aciduria
  Ornithine-α-ketoacid
    transaminase deficiency

Carbohydrate disorders
  Glycogen storage disease
    (types II, III and IV)
  Galactosaemia
  Mannosidosis
  G6PD deficiency

Lipid disorders
  Gaucher's disease
  Generalized gangliosidosis
  Beta-galactosidase deficiency
    (Tay-Sachs disease)
  Sandhoff's disease
  Metachromatic leukodystrophy
  Niemann-Pick disease
  Refsum's disease

Mucopolysaccharidoses
  Hunter's syndrome
  Hurler's syndrome

Miscellaneous traits
  Adrenogenital syndrome
  Lesch-Nyhan syndrome
  Lysomal acid phosphatase
    deficiency
  Xeroderma pigmentosum
  Acatalasaemia
  Chediak-Higashi syndrome
  Congenital erythropoietic
    porphyria
  Cystic fibrosis
  I-cell disease
  Oroticaciduria
  Sickle-cell disease

Chromosomal aberrations
  Many: about 1 in every 200 live
    births has one

Index